Nutrient Management Strategies in Integrated Farming Systems

Nutrient Management Strategies in Integrated Farming Systems

Editor

Neelima Watve

scitus academics

Nutrient Management Strategies in Integrated Farming Systems

Edited by **Neelima Watve**

Printed in 2017

ISBN: 978-1-68117-082-4

Library of Congress Control Number: 2015935480

SCITUS Academics LLC,
616, Corporate Way, Suite 2, 4766,
Valley Cottage, NY 10989

www.scitusacademics.com

Contents

Preface

The overall strategy for increasing crop yields and sustaining them at a high level must include an integrated approach to the management of soil nutrients, along with other complementary measures. An integrated approach recognizes that soils are the storehouse of most of the plant nutrients essential for plant growth and that the way in which nutrients are managed will have a major impact on plant growth, soil fertility, and agricultural sustainability. Farmers, researchers, institutions, and government all have an important role to play in sustaining agricultural productivity. Concerns also are growing about the long-term sustainability of agriculture. Both the over- and under application of fertilizer and the poor management of resources have damaged the environment. In developed countries, for example, over application of inorganic and organic fertilizer1 has led to environmental contamination of water supplies and soils. In developing countries, harsh climatic conditions, population pressure, land constraints, and the decline of traditional soil management practices have often reduced soil fertility. Because agriculture is a soil-based industry that extracts nutrients from the soil, effective and efficient approaches to slowing that removal and returning nutrients to the soil will be required in order to maintain and increase crop productivity and sustain agriculture for the long term.

Editor

Chapter 1

Soil Health Management under Hill Agroecosystem of North East India

R. Saha, R. S. Chaudhary, and J. Somasundaram

Indian Institute of Soil Science, Indian Council of Agricultural Research, Nabibagh, Berasia Road, Bhopal, Madhya Pradesh 462 038, India

ABSTRACT

The deterioration of soil quality/health is the combined result of soil fertility, biological degradation (decline of organic matter, biomass C, decrease in activity and diversity of soil fauna), increase in erodibility, acidity, and salinity, and exposure of compact subsoil of poor physicochemical properties. Northeast India is characterized by high soil acidity/Al^{+3}toxicity, heavy soil, and carbon loss, severe water scarcity during most parts of year though it is known as high rainfall area. The extent of soil and nutrient transfer, causing environmental degradation in North eastern India, has been estimated to be about 601 million tons of soil, and 685.8, 99.8, 511.1, 22.6, 14.0, 57.1, and 43.0 thousand tones of N, P, K, Mn, Zn, Ca, and Mg, respectively. Excessive deforestation coupled with shifting cultivation practices have resulted in tremendous soil loss (200 t/ha/yr), poor soil physical health in this region. Studies on soil erodibility characteristics under various land use systems in Northeastern Hill (NEH) Region depicted that shifting cultivation had the highest erosion ratio (12.46) and soil loss (30.2–170.2 t/ha/yr), followed by conventional agriculture system (10.42 and 5.10–68.20 t/ha/yr, resp.). The challenge before us is to maintain equilibrium between resources and their use to have a stable ecosystem. Agroforestry systems like agri-horti-silvi-pastoral system performed better over shifting cultivation in terms of improvement in soil organic carbon; SOC (44.8%), mean weight diameter; MWD (29.4%), dispersion ratio (52.9%), soil loss (99.3%), soil erosion ratio (45.9%), and in-situsoil moisture conservation (20.6%) under the high rainfall, moderate to steep slopes, and shallow soil depth conditions. Multipurpose trees (MPTs) also played an important role on soil rejuvenation. Michelia oblonga is reported to be a better choice as bioameliorant for these soils as continuous leaf litter and root exudates improved soil physical behaviour and SOC considerably. Considering the present level of resource degradation, some resource conservation techniques like zero tillage/minimum tillage, hedge crop, mulching, cover crop need due attention for building up of organic matter status for sustaining soil health.

INTRODUCTION

Soil degradation has raised some serious debate, and it is an important issue in the modern era. It refers to the decline in soil's inherent capacity to produce economic goods and perform ecologic functions. It is the net result of dynamic soil degradative and restorative processes regulated by natural and anthropogenic factors. The degree of soil degradation depends on soil's susceptibility to degradative processes, land use, the duration of degradative land use, and the management. Soil and water degradation are also related to overall environmental quality, of which water pollution and the "greenhouse effect" are two major concerns of global significance. Recent global concerns over increased atmospheric CO_2, which can potentially alter the earth's climate systems, have resulted in raising interest in studying Soil organic matter (SOM) dynamics and carbon (SOC) sequestration capacity in various ecosystems [1]. Soils represent an important terrestrial stock of

C and approximately two to three times as much as terrestrial vegetation and atmosphere, respectively, and the C in the SOM of agricultural land is composed of dominant terrestrial C stock. Soil quality is the capacity of a soil to function within ecosystem boundaries to sustain biological productivity, maintain environmental quality, and promote plant and animal health and thus has a profound effect on the health and productivity of a given ecosystem and the environment related to it.

The North Eastern parts of India, comprising the states of Arunachal Pradesh, Assam, Manipur, Meghalaya, Mizoram, Nagaland, Sikkim, and Tripura, lies between 22 05′ and 29 30′ N latitudes and 87 55′ and 97 24′ E longitudes. The region is characterised by diverse agroclimatic and geographical situations. About 54.1 percent of the total geographical area is under forests, 16.6 percent under crops, and the rest either under nonagricultural uses or uncultivated land. The low area under agricultural crops is due to natural corollary of the physiographic features of the region, as major chunk of the land has more than 15 percent slope, undulating topography, highly eroded and degraded soils, and inaccessible terrain. Continuous dilution of the forest cover in the region due to shifting cultivation, firewood, and timber collection is posing the most crucial problem resulting in poor soil health and environmental degradation in the hills.

SHIFTING CULTIVATION

Shifting cultivation, also known as Jhum cultivation, is the most traditional and dominant land use system in this region. On an average, 3,869 km^2 area is put under shifting cultivation every year. Shifting cultivation in its more traditional and cultural integrated form is an ecological and economically viable system of agriculture as long as population densities are low and jhum cycles are long enough to maintain soil fertility. The system involves cultivation of crops in steep slopes. Land is cleared by cutting of forests, bushes, and so forth up to the stump level, leaving the cut materials for drying and finally burning to make the land ready for sowing of seeds of different crops before the onset of rains. The cultivation is confined to a village boundary and often after two or three years, the cultivated area is abandoned and a new site is selected to repeat the process. The shifting cultivation became unsustainable today primarily due to the increase in population that led to increase in food demand. Jhuming cycle in the same land, which extended to 20–30 years in earlier days, has now been reduced to 3–6 years [2]. Land degradation in the region is 36.64% of the total geographical area, which is almost double than the national average of 20.17% [3]. The problem of land degradation is much serious in the states like Manipur, Nagaland, and Sikkim, where more than 50% of total geographical area is defined as wastelands. Of various degradation types, water erosion, reduced infiltration, acidification, nutrient leaching, burning of vegetation, decline in vegetative cover, and biodiversity are important in context to the NE region.

EFFECT OF SHIFTING CULTIVATION

Change in Forest Cover

The total forest cover in the region is 1,41,652 km^2, which is about 54.1% of the geographic area as against the national average of 19.39% [4]. Manipur and Meghalaya have dense forest cover of 25.57 and 25.33%, respectively (Table 1). Similarly for Nagaland, Sikkim, Tripura, and Mizoram, the dense forest cover is 32.53, 33.70, 33.02 and 42.39%, respectively. Among seven sisters of NEH, Arunachal Pradesh is the only state, which has the dense forest cover of 64.0%. Since shifting cultivation is still practiced in the region, and every year dense forest is converted intojhum fields, there is drastic reduction in dense forest cover (canopy density > 40%) in most of the states.

Table 1: Trends of forest loss/gain (km^2) in NEH region

States	1999 Assessment			2001 Assessment			Net difference
	Dense forest cover	Open forest cover	Total forest cover	Dense forest cover	Open forest cover	Total forest cover	
Arunachal Pradesh	57,756	11,091	68,847	53,932	14,113	68,045	(−) 802
*Assam	15,548	8,276	23,824	14,517	9,171	23,688	(−)136
Manipur	5,936	11,448	17,384	5,710	11,216	16,926	(−) 458
Meghalaya	5,925	9,708	15,633	5,681	9,903	15,584	(−) 49
Mizoram	3,786	14,552	18,338	8,936	8,558	17,494	(−) 844
Nagaland	5,137	9,027	14,164	5,393	7,952	13,345	(−) 819
Sikkim	2,363	755	3,118	2,391	802	3,193	75
Tripura	2,228	3,517	5,745	3,463	3,602	7,065	1320
**Total	83,131	60,098	1,43,229	85,506	56,146	1,41,652	(−) 1577

*Data for Assam is during the assessment year of 1997–1999 and **total reports only for NEH region.

Source: [4].

Effect of Burning on Soil Fertility

The burning process related to shifting cultivation practices has tremendous effect on soil ecosystem. The impact of fire on ecosystem is profound and its consequences are dependent on intensity and frequency of fire, proportion of biomass burned, the time of monsoon setting, and total annual precipitation. The extent to which organic matter is transformed into ash depends on a number of factors viz intensity and duration of fire, fuel load, moisture content in the fuel, weather, and topography. Burning of above-ground vegetation showed an increase in pH and cations and a decrease in carbon and nitrogen contents in the surface soil [5]. Quick release of nutrients especially cations after burning has been reported by Kellman et al. [6]. The organic carbon content of soil decreased drastically after burning because of oxidation loss. Rise in pH, temperature, and bases of the soil might have increased the microbial activity after burning which in turn resulted in accelerating mineralization of organic N to inorganic forms [7, 8].

Soil Erosion and Nutrient Loss

Soil erosion under shifting cultivation is highly erratic from year to year depending on rainfall characteristics. Studies on steep slopes (44–53%) have indicated the soil loss to the tune of 40.9 t/ha and the corresponding nutrient losses per hectare are 702.9 kg of organic carbon, 63.5 kg of P and 5.9 kg of K [9]. The soil loss from hill slopes (60–79%) under first year, second year, and abandoned jhum was estimated to be 147, 170, and 30 t/ha/yr [10]. In general, tolerable soil loss (T) value is 11.2 Mg/ha/yr (5.0 t/ac/yr) while it is between 5.0 and 12.5 Mg/ha/yr (2.2 and 5.6 t/ac/yr) in North West Himalayas [11]. During first few years of clearing, carbon and nitrogen levels decrease rapidly. According to one estimate annual loss of top soil, N, P and K due to shifting cultivation is 88346, 10669, 0.372, and 6051 thousand tones in the region [12]. Singh et al.[13] reported nutrient loss to the tune of 6.0 million tones of organic carbon, 9.7 tones of available P, and 5690 tones of K from the NEH region. Nutrient losses from the jhum field through runoff and percolation are rather heavy during cropping.

LONG-TERM STRATEGIES FOR RESOURCE CONSERVATION AND IMPROVEMENT IN SOIL HEALTH

Nearly 37.1% of the total geographical area in Northeast India is under the threat of land degradation, where erosion is a major land degradative process. With the great concern of poor soil health and severe land degradation, there is a need of viable option for ecorestoration and maintenance of soil resources which could sustain long-term soil productivity and improve food security of the poor tribal farmers of northeast India under the humid subtropical climate of the north-eastern Himalayan region. Three broad strategies, suitable for different land situation, elevation, and topography prevailing in this region, are discussed here.

Multipurpose Trees (MPTs)

The multipurpose tree species (MPTs) form an integral component of different agroforestry interventions in crop sustainability. The MPTs, besides furnishing the multiple outputs like fuel, fodder, timber, and other miscellaneous products, help in improvement of soil health and other ecological conditions. Farmers of the region integrate various tree species in different land use in the region; however, priority species vary from state to state and even from place to place within a state based on ethnic diversity and food habits of the tribal communities. In the region, as many as 40 promising species are cultivated in tropical and subtropical region, and 30 in temperate zone of the region in different farming systems. Besides, 28 bamboo species and 2 genera of cane also find a place in various agroforestry programmes. Tree density ha^{-1} is also a crucial factor on sloppy lands. In general, optimum tree density in case of agri-horticulture system is 400 trees/ha, while in agri-silviculture, it is 200 plants/ha so as to minimize the effect of shade and biochemical interactions on growth and production of agricultural crops [14, 15].

Long-term effect of various multipurpose tree species on soil physical behaviour has been studied [16]. Multipurpose tree species with greater surface cover, constant leaf litter fall, and extensive root system increased soil organic C by 96.2%, porosity by 10.9%, aggregate stability by 24.0%, and available soil moisture by 33.2% and simultaneously reduced bulk density and erosion ratio by 15.9 and 39.5%, respectively (Table 2). Among the tree species tested, P. kesiya, M. oblonga and Alnus nepalensis were found suitable as bioameliorant in hilly terrain of northeast India in terms of organic matter buildup through presence of leaf litter, better soil aggregation, transmissivity, and infiltrability through extensive root system, improved soil conservation through constant surface cover with leaf biomass. Such improvement in soil hydrophysical properties in tree-based system has a direct bearing on long-term sustainability, productivity, and soil quality in hilly ecosystem.

Table 2: Effect of various MPTs on soil physical properties

MPTs	Organic C (g kg^{-1})	Aggregate stability	Available water (m^3 m^{-3})	Infiltration rate (mm h^{-1})	Erosion ratio
Pinus kesiya	35.4	75.6	0.220	8.04	0.20
Alnus nepalensis	32.2	72.1	0.201	7.28	0.23
Parkia roxburghii	23.1	63.4	0.192	4.85	0.30
Michelia oblonga	33.6	73.2	0.210	6.10	0.22
Gmelina arboria	28.6	67.9	0.183	5.36	0.24
Control (No tree)	15.6	56.8	0.151	3.84	0.39

Agroforestry Interventions in Degraded Lands

The region has a very high rate of land degradation. In this region, 7.85 million ha area is degraded which need rehabilitation through various agroforestry models [3]. Agroforestry system (AFS) has today become an established approach of integrated land management system not only for renewable resource production but also for ecological consideration. It represents the integration of agriculture and forestry to increase the productivity and sustainability of farming system.

Soil Fertility Buildup

Study revealed that organic carbon, available P, and exchangeable cations contents in surface soil ranged in between 2.0–2.5%, 10.4–13.2 ppm, and 5.9–8.4 cmol (p^+) kg^{-1}, respectively, under jackfruit-based AFS, while 1.5–1.8% organic carbon, 3.8–6.7 ppm available P, and 3.9–5.9 cmol (p^+) kg^{-1} total cations were found under arecanut/khasi mandarin-based AFS [17].

In an another study, long-term effect of agri-horticulture (comprising Khasi mandarin + agricultural crops, and Assam lemon + agricultural crops), agri-silviculture (multipurpose tree species + annual agricultural crops), silvi-horti-pastoral (alder + pine apple + fodder grasses), and multistoried AFS (alder + tea + black pepper + annual agricultural crops between the tree rows) on soil properties and fertility status was evaluated in acid Alfisol of Meghalaya compared with natural forest as a control. In all the AFS, significant (1.17–1.65 fold) increase in organic carbon was found as compared to initial status, the maximum contribution being by silvi-horti-pastoral AFS. The same system also registered 43.2% higher exchangeable Al compared to natural forest and consequently a maximum decrease of 0.50 units in pH (Table 3). The exchangeable Ca, Mg, Na, K, and Al and available N, P, and K content were higher in all the systems compared to natural forest and the content of these nutrients decreased with increasing soil depth [18].

Table 3: Effect of agroforestry systems on soil properties

Soil properties	Agroforestry systems					
	Agrisilviculture	**Agrihorti culture (khasi mandarin + crops)**	**Agrihorti culture (Assam lemon + crops)**	**Silvihorti pastoral (Alder + pine apple + fodder grass)**	**Multistoried AFS (Alder + tea + black pepper + crops)**	**Natural forest**
pH	4.65	4.62	4.80	4.25	4.61	4.62
Organic C (%)	1.62	1.55	2.02	2.19	1.91	1.92
Exchangeable Ca [cmol (p^+) kg^{-1}]	0.40	0.86	0.74	0.31	0.65	0.26
Exchangeable Mg [cmol (p^+) kg^{-1}]	0.75	0.51	0.33	0.48	0.71	0.16
Exchangeable K [cmol (p^+) kg^{-1}]	0.232	0.244	0.249	0.238	0.201	0.169
Exchangeable Na [cmol (p^+) kg^{-1}]	0.201	0.220	0.194	0.195	0.197	0.196
Exchangeable Al [cmol (p^+) kg^{-1}]	2.65	2.70	2.20	3.15	2.05	2.20
Available N (ppm)	190.1	180.8	203.6	199.4	216.9	167.2
Available P (ppm)	2.75	4.10	5.36	0.94	3.36	0.63
Available Fe (ppm)	8.9	10.4	12.8	10.9	13.9	7.3
Available Mn (ppm)	0.58	0.92	0.79	0.83	1.04	0.04
Available Zn (ppm)	0.08	0.05	0.07	0.006	0.08	0.025
Available Cu (ppm)	0.21	0.23	0.37	0.30	0.27	0.10

Source: [18].

In a study under Farming System Research Project (FSRP) carried out at ICAR Research Complex, Barapani, effect of various AFS like silvi-pastoral, silvi-horticulture, agri-horti-silvipastoral has been evaluated [19] after 17 years of their adoption on soil fertility indices (Table4). The natural fallow and abandoned jhum land at Umiam were taken for comparison. Organic carbon content increased in all the AFS including natural fallow, however, the quantity largely depended on the nature of vegetation in different systems. Adoption of different cropping pattern in various AFS markedly influenced the exchangeable Ca, Mg, and K content in the soil. Maximum accumulation of these cations was recorded under agri-horti-silvipastoral and silvi-horticulture AFS followed by natural fallow and silvi-pastoral systems. Accumulation of exchangeable K was maximum in silvi-horticulture followed by agri-horti-silvipastoral. The available N, P, and S were higher in agri-horti-silvipastoral and silvi-horticulture compared to natural fallow and silvi-pastoral AFS.

Table 4: Effect of different land use systems developed under FSRP, Meghalaya on soil properties.

Characteristics	Agroforestry systems				
	Natural fallow	**Abandoned jhumland**	**Silvipastoral**	**Agri-horti-silvipastoral**	**Silvihorti-culture**
pH	4.99 (4.90)	4.76 (5.20)	4.52 (5.10)	4.92 (4.90)	4.91 (4.90)
Organic C (%)	2.94 (1.85)	3.42 (1.90)	2.61 (1.80)	2.97 (1.82)	2.97 (1.80)
Exchangeable Ca [cmol (p^+) kg^{-1}]	1.96 (1.15)	1.57 (1.16)	1.25 (1.10)	2.11 (1.20)	2.00 (1.20)
Exchangeable Mg [cmol (p^+) kg^{-1}]	0.55 (1.15)	0.38 (1.16)	0.43 (1.20)	1.15 (1.20)	0.85 (0.60)
Exchangeable Al [cmol (p^+) kg^{-1}]	0.88	1.30	1.56	0.90	0.90
Available N (ppm)	179.2	251.1	214.5	220.3	210.9
Available P (ppm)	1.9	2.0	2.1	16.6	12.9
Available K (ppm)	175.6	130.8	98.0	162.7	265.0
Available S (ppm)	14.5	14.8	10.4	19.9	12.9

Figures in parentheses indicate the initial values at the start of the project.

Source: [19].

Soil Physical Health

Effect of various land use systems on soil physical properties shown in Table 5 indicated that the maximum reduction in bulk density over shifting cultivation was recorded in forest (17.6%) followed by agri-horti-silvi-pastoral (14.3%), livestock based (13.4%), natural fallow, and agriculture system (12.6%). Higher percentage of macroaggregates (54.5%), organic C content (2.95%), and biotic activity were also observed in forest ecosystem. Soil biota influences soil properties through formation of stable aggregates, development of organomineral complexes by improving macroporosity and continuity of pores from surface to the subsoil which ultimately increase the water transmission and reduce run-off. Higher transmission and storage pore volume coupled with lower value of residual pores associated with modified land use systems as compared to shifting-cultivated plots was thus an indication of maintaining the pore geometry of the soil under these systems. The better soil aggregation under natural forest, multistoried AFS, and silvihortipastoral systems maintaining intensive vegetative cover throughout the year could be ascribed to the effect of higher percentage of organic matter, clay content, and high amount of Al and Fe oxides in soil.

Table 5: Effect of different land use systems developed under FSRP, Meghalaya on soil physical properties

Soil properties	Land use systems					
	Agriculture	**Agri-horti-silvipastoral**	**Forestry**	**Livestock based**	**Natural fallow**	**Shifting cultivation**
Bulk density (Mg m^{-3})	1.04	1.02	0.98	1.03	1.04	1.19
Total Porosity (%)	59.67	60.47	62.02	60.08	66.23	53.88
Macroaggregates (>0.25 mm)	21.72	54.19	54.47	50.02	50.53	18.17
Microaggregates (<0.25 mm)	47.85	23.23	23.81	22.80	21.90	42.34
MWD (mm)	2.76	2.99	3.16	2.85	2.93	2.31
Available water (m^3 m^{-3})	0.210	0.222	0.231	0.220	0.233	0.169
Hydraulic Conductivity (cm hr^{-1})	2.74	4.72	5.47	2.95	6.66	2.09

Source: [20].

Soil and Water Conservation

Some of the potential farming systems such as agriculture on bench terraces, horticulture, and agri-horti-silvipastoral systems have been evaluated [21] at the experimental watershed of ICAR Research Complex at Barapani for long-term runoff, soil and nutrient losses, production behaviour, biotic and abiotic changes, and so on. The data indicated that mixed land use systems with appropriate soil conservation measures, namely, bench terraces, contour trenches, and so forth, were the most effective in retaining 90–100% annual rainfall and simulated the effects of natural forest. The contributions to stream flow in the watersheds having substantial area under natural forest is primarily by subsurface flow (base flow). The watersheds having continuous stream flow characteristics generated base flow to the extent of 70–90% of its total water yields. As expected, the watershed treated with jhum (shifting) cultivation yielded the highest peak runoff while the one left undisturbed with natural vegetation gave the minimum peak runoff. The results revealed that agroforestry and other mixed land use systems most effectively conserved moisture and substantially reduced peak runoff (Tables 6(a) and 6(b)). The low erosion ratio values in silvi-horti-pastoral and multistoried AFS (3.07 and 3.06, resp.) showed that these systems were the most suitable for soil and water conservation in hilly ecosystem [22]. This could be ascribed to the effect of heavy litter fall, which might have increased the cohesiveness in the soil system after decomposition and also binds the soil tightly in lower horizons by their deep root systems.

Table 6: (a) Pretreatment (year) precipitation, storm flow, peak flow rate in different land use systems. (b) posttreatment water yield, base flow, and peak flow in different land use systems (average of nine years).

(a)

Land use system	Precipitation (mm)	Threshold rainfall (mm)	Total water yield (mm)	Total water yield (% of rainfall)	Surface runoff (mm)	Peak flow (mm hr^{-1})
Dairy based farming	2249.30	363.20	27.21	1.20	27.21	3909
Forestry block	2249.30	399.90	655.21	29.12	54.30	16.94
Agroforestry	2249.30	533.70	32.55	1.45	9.90	6.17
Agropastoral	2249.30	364.30	25.50	1.13	25.50	31.86
Agrohortisilvipastoral	2249.30	348.60	4.10	0.18	4.10	10.45
Natural fallow	2249.30	541.70	2.87	0.13	2.87	13.65
Shifting cultivation	2249.30	1634.5	15.88	0.70	15.88	35.30

(b)

Land use systems	Annual water yield range (mm)	Mean water yield (mm)	Mean water yield (% of annual rainfall)	Maximum peak flow (mm hr^{-1})
Dairy based farming	0–66.699	9.56	0.37	7.81
Forestry block	67.42–1013.88	371.90	4.73	13.54
Agroforestry	39.31–648.26	241.14	9.55	12.87
Agropastoral	0.60–62.49	12.47	0.69	20.71
Agrohortisilvipastoral	0.24–121.91	28.98	1.14	12.07
Natural fallow	0–51.39	11.77	0.46	4.49
Shifting cultivation	0–517.72	102.94	4.07	86.10

Source: [21].

Soil C Sequestration Potential

Assessment of soil quality is an invaluable tool in determining the sustainability and environmental impact of agricultural ecosystems. Soil quality under different agroecosystems using soil organic carbon (SOC) and soil microbial C (SMBC) as soil quality indicators suggests that the shifting cultivated areas had the lowest SMBC value of 192 mg/kg while soil underMichelia oblonga plantation had the significantly () highest value of 478 mg/kg. The proportion of SMBC to total soil organic carbon (SOC) was in the range of 0.76 to 1.96% across all the systems. Multipurpose tree species like P. kesiya, A. nepalensis, P. roxburghii, M. oblonga, andG. arboria with greater surface cover, constant leaf litter fall, and extensive root systems increased soil organic carbon by 96.2% (Table 7), helped with better aggregate stability by 24.0%, improved available soil moisture by 33.2%, and in turn reduced soil erosion by 39.5% [16, 23]. Similarly, a comparative study on the effect of various MPTs on soil organic carbon pool (Table 8) showed a concomitant rise in SOC in soils under MPTs and a subsequent decline in soils of open space over 4–16 years. Maximum rise in SOC was noticed in soils of A. indica (28.6 Mg/hm^2) followed by A. Aurculiformisi (21.9 Mg/hm^2), G. arborea (21.8 Mg/hm^2), M. Champaca (16.7 Mg/hm^2), and so forth. The minimum rise in SOC was noted in soils under T. grandis. So an increase of SOC was noted from 3.8 Mg/hm^2 in soils of open space to 19.5 Mg/hm^2 in that under MPTs after 16 years. The comparatively high humin carbon present in soils under A. auriculiformis, L. leucocephala,and G. Arborea indicated the enhanced storage of organic carbon pool in agroforestry systems [24]. Swamy et al. [25] estimated that a six-year-old G. arborea, based agri-sivicultural systems in India sequestered 31.4 Mg hm^{-2} carbon.

Table 7: Growth, litter production, fine root biomass of promising MPTs in humid tropics, and their contribution on SOC content

MPT	Annual litter production (g m^{-2})	Time required for decomposition (days)	Total fine root biomass (g m^{-2})	Organic C (g kg^{-1})
P. kesiya	621.5	718	496.75	35.4
A. nepalensis	473.75	350	435.50	32.2
P. roxburghii	341.75	385	415.50	23.1
M. oblonga	512.25	390	462.00	33.6
G. arboria	431.75	360	419.00	28.6

Source: [16].

Table 8: Changes in SOC (Mg hm^{-2}) over the years under various MPTs in humid tropics

MPTs	Years			
	4	**8**	**12**	**16**
A. auriculiformis	11.1	11.9	17.9	21.9
M. alba	9.9	9.9	9.9	15.9
L. leucocephala	11.5	11.5	12.8	16.7
D. sissoo	13.1	12.5	13.1	13.9
G. maculate	13.1	13.1	13.9	14.9
A. indica	10.9	10.9	14.7	28.6
M. champaca	13.9	13.7	13.9	16.9
E. hybrid	9.9	9.9	14.9	16.1
T. grandis	11.5	11.3	11.5	12.9
G. arborea	12.2	12.2	12.8	21.8
S. saman	10.6	11.3	11.3	13.9
A. procera	13.5	13.1	13.5	14.7
Open space (Control)	11.9	11.9	11.1	9.1

Source: [24].

Resource Conservation Techniques

Conservation Tillage

Conservation tillage are system of managing crop residue on the soil surface with minimum disturbance. The stubble mulch or reduced tillage/minimum tillage, no tillage and direct drill are components of conservation tillage. The objectives are (i) to leave enough plant residue on the soil surface at all times for water, and wind erosion control, (ii) to conserve soil and water and (iii) to reduce energy use [26]. Some of the conservation tillage practices followed in hill ecosystems are discussed here.

In-Situ Residue Management

Low native soil nitrogen (N) and very low phosphorus (P) coupled with apathy of farmers towards use of fertilizer is the major constraints limiting the rice productivity in NEH Region of India. Productivity and nutrient recycling potential in rice (Oryza sativa L.)—vegetables cropping sequences under low input in-situ residue management under rainfed condition was evaluated on lowland situation at ICAR Research Complex for NEH Region, Umiam, Meghalaya. After harvesting of rice, five vegetable crops, viz., tomato, potato, frenchbean, cabbage, and carrot, were grown. No external input including fertilizer, pesticides, and so forth was applied except one hand weeding at 30 days after transplanting in case of rice. In case of vegetables, only one earthing up and intercultural operations were done as per the requirement of the crops. Only the economic parts of crops were harvested and left-out portion including weed residues were chopped and incorporated into the soil. A considerable amount of nutrients were recycled through in-situ weed biomass incorporation. The weed biomass ranged from 37.5 q/ha with rice-tomato to 50.6 q/ha in rice-fallow. Highest amount of NPK recycling was recorded from rice-potato sequence. Soil fertility in terms of available NPK status analysed after 4 years was found stable in all the crop sequences except rice-cabbage, where it declined slightly. The soil biological properties like population of Rhizobium, bacteria, phosphorus solubilizing microorganisms, and earthworm activity all were found remarkably higher in experimental field compared to plots that are managed inorganically.

Incorporation of Jungle Grass

Long-term effects of different locally available grasses and weeds on soil hydro-physical properties and rice yield through a 5-year field experimentation under hilly ecosystem of Meghalaya depicted that incorporation of jungle grass (Ambrosia spp.), in puddled rice soil improved soil organic carbon (SOC) by 21.1%, the stability of microaggregates, moisture retention capacity, and infiltration rate of the soil by 82.5, 10, and 31.3%, respectively, and soil bulk density decreased by 12.6% [27]. Locally available jungle grasses are equally good as an organic amendment, which would also ease the problem of disposal of these grasses during peak monsoon. Therefore, these organic sources may serve as alternative to farm yard manure (FYM) and have a dramatic effect on long-term productivity of rice.

Zero Tillage

Zero tillage in rice-based system improves physical properties of soil like soil structure, increase the relative proportion of biochannels, macropores, and decrease the susceptibility of crusting. It has been observed that the bulk density of soil decreased about 25%, total porosity and soil aggregates increased by 29 and 32%, respectively, over the conventional tillage practices (2-3 passes of powertiller/spade). It also increases the SOC content by 12.5%, available P by 14.3%, and K by 29.4% over conventional tillage. Zero tillage saved 20% energy (Figure 1) and fertilizer needs as compared other conventional tillage methods by conserving soil and water [28] without jeopardizing the crop production (rice yield of 37 q/ha). In other tillage practices like power-tilled, desiploughed, or manually weeding, the energy in terms of labour requirement was much higher.

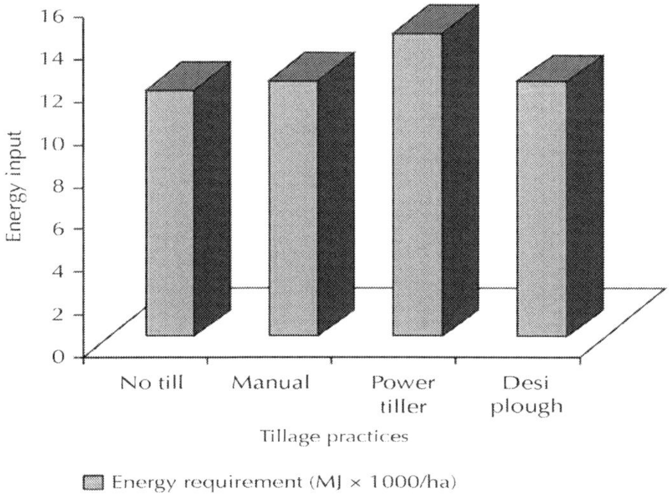

Figure 1: Energy requirement of different tillage practices. Source: [28].

Integrated Plant Nutrient Supply

Integrated use of balanced inorganic fertilizer in combination with lime and organic manure sustains a better soil health for achieving higher crop productivity under intensive cropping systems in hilly ecosystem of north eastern India. Study suggests that addition of NPK fertilizers along with organic manure, lime, and biofertilizers had increased SOC content, aggregate stability, moisture retention capacity, and infiltration rate of the soil while reducing bulk density. The SOC content under the treatment 100% NPK + lime + biofertilizer + FYM was significantly higher (68.6%) than control plots [29].

Pastural Development

Resource conserving and environmental friendly production strategies are desirable for agrarian economies. Grass cover is the key factor in improving soil physicochemical health by assuring regular addition of organic matter, thus reducing surface runoff and soil erosion. Some promising perennial grasses like Setaria, Congosignal, Guinea, Napier, and Broom grass were tested for their effect on soil physicochemical properties. Study [30] revealed that continuous 15 years grass covers significantly increased the SOC, the highest SOC content with Setaria (2.24%). Similarly, Soil microbial biomass carbon, soil aggregation, and infiltration rate under various grass covers were also high as compared to plots without grass covers.

Hedgerow Intercropping

As the trees have long gestation period, farmers may be reluctant to cultivate the trees mainly due to prevailing land tenure system in the region. However, cultivating various hedgerow species even in jhum field could be better option for them as these species have short gestation period. Hedgerows alone reduced soil loss by 94% and run-off by 78%. When twigs and tender stem of hedge plants are used for mulch, it conserved 83% of the soil and 42% of rainfall. In a study conducted at Changki, Nagaland in NEH region, the soil loss was reduced by 22% with the incorporation of hedgerow species in the jhum fields compared to traditional jhum site (38.14 t/ha/yr). Thus contour hedgerow technology provides an option for farming on the hill slopes on a sustainable basis. Growing of nitrogen fixing hedge species on the field bunds helps in fixation of atmospheric nitrogen and reduces the leaching losses of mineral nitrogen. Their vigorous root system mobilizes phosphorus, potassium, and other trace elements. ICAR Research Complex for NEH Region has also screened various hedgerow species for plantation, and Cajanus cajan, Crotalaria tetragona, Desmodium rensonii, Flemingia macrophylla, Indigofera tinctoria, Tephrosia candida, and Gliricidia maculata have been found suitable for farming in Eastern Himalayas. Survival percentage of these species ranged from 60.0 to 80.0 over degraded sites. The total N, P, and K concentration in the foliage of hedgerow species ranged from 3.23–3.86; 0.32–0.81; 1.26–1.67%, respectively. Total leaf biomass production on the dry weight basis after one year of growth was found to be highest in C. tetragona (22.98 q/ha) followed by G. maculata (20.75 q/ha), I. tinctoria (16.99 q/ha), and T. candida (15.30 q/ha). Among the hedgerow species, C. tetragona enriched the soil fertility more efficiently as it accumulated higher amount of total N, P, and K (79.74, 11.03, and 37.46 kg/ha) through its leaf incorporation. The recycling of bases in litter of hedgerow could potentially counteract the acidification [31]. The incorporation of leaf biomass of T. candida improved the pH in acid soil by increasing 0.49 units from the initial level at surface soil. Thus, the biomass produced from hedgerows showed a favorable influence on soil acidity.

Organic Farming

Organic farming is primarily in operation in areas under shifting cultivation and traditional land use systems in north east India. Nearly 57.1% of total geographical area (TGA) in India is under the threat of land degradation mainly by water erosion. On an average, 37.1% of TGA in NE India is in degraded state. Fertilizer use in most of the states of the region is far below the national average. The use of N, P, and K through fertilizer in the region is only 13.37, 11.12, and 11.0% of the crop removal thus necessasiating the organic source of nutrition in the domain of soil health management. Organic sources if pooled together can supply 13.07 kg N/ha, 7.18 kg phosphate/ha, and 7.34 kg potash/ha in NE India. The micronutrient supply from organic sources may be adequate. Substantial amount of potash can be obtained from crop residues if managed to add in soils. Biofertilizers in case of adequate supply can produce an increase (5–30%) in yield. Vermicomposting of rural wastes holds a great promise in mitigating nutrient hunger of soils in NE India considering supply of composting earthworms and need based training in compost technology. Soil amelioration with the use of limestone deposit available in north east can be brought in use. Finally, watershed based technology with proper soil and water conservation measures can be an effective avenue to nurture soil health for sustainable organic food production.

EPILOGUE

Even today, Jhuming is considered as a major source of rural economy in north eastern part of India and will remain as important one as it is associated with socioeconomic and cultural systems of the people of this region. Because of this, degradation will continue in the years to come and may reach to the extent of out of control, if proper care is not taken right now. Therefore, to reduce all types of degradation level, a comprehensive forest policy is required as a long-term strategy in the region for sustainability and augmentation of food, fuel, fodder, and timer requirements. In this direction, agroforestry coupled with some sound resource conservation techniques needs to be strengthened for long-term sustainable production and environmental conservations in fragile ecosystem which will contribute to improved food security and income generation for resource poor farmers and protect the environments.

Integrated farming system (IFS) has emerged as a well accepted, single window, and sound strategy for harmonizing simultaneously jointmanagement of land, water, vegetation, livestock, and human resources, The IFS developed for hill areas could reduce the risk of soil degradation, produce the soils productive potential, and reduce the risks of environmental degradation. Besides, these interventions having a tree crop with a high quality of leaf litter and root binding ability reduce erodibility of rainfall/runoff and improve the physicochemical conditions. Attempt should also be made to manage soil health through addition of organic inputs in this region.

REFERENCES

1. W. H. Schlesinger, "Carbon sequestration in soils," Science, vol. 284, no. 5423, p. 2095, 1999.
2. D. N. Borthakur, Agriculture of the North Eastern Region with Special Reference to Hill Agriculture, Beecee Prakashan, Guwahati, India, 1992.
3. Anonymous, Wastelands Atlas of India, Ministry of Rural Development, Government of India and National Remote Sensing Agency, Hyderabad, India, 2000.
4. Anonymous, State of Forest Report, Forest Survey of India. Ministry of Environment and Forests, Government of India, 2001.
5. S. C. Ram and P. S. Ramakrishnan, "Hydrology and soil fertility of degraded grasslands at Cherrapunji in North Eastern India," Environment Conservation, vol. 15, pp. 29–35, 1988.
6. M. Kellman, K. Miyanishi, and P. Hiebert, "Nutrient retention by savanna ecosystems II. Retention after fire," Journal of Ecology, vol. 73, no. 3, pp. 953–962, 1985.

7. I. F. Ahlgren and C. E. Ahlgren, "Effect of prescribed burning on soil microorganisms in a Minnesota jack pine forest," Ecology, vol. 46, no. 3, pp. 304–310, 1965.
8. G. Griffith, "Fertility problems in Uganda," Technical Communication, Commonwealth Bureau of Soil Science, vol. 46, pp. 160–164, 1949.
9. M. Ram and B. P. Singh, "Soil fertility management in farming systems," Lectures notes, off campus training on farming system, Aizawl, India, 1993.
10. A. Singh and M. D. Singh, "Effect of various stages of shifting cultivation on soil erosion from steep hill slopes," Indian Forester, vol. 106, no. 2, pp. 115–121, 1981.
11. D. Mandal, K. S. Dadhwal, O. P. S. Khola, and B. L. Dhyani, "Adjusted T values for conservation planning in Northwest Himalayas of India," Journal of Soil and Water Conservation, vol. 61, no. 6, pp. 391–397, 2006.
12. U. C. Sharma, "Methods of selecting suitable land use system with reference to shifting cultivation in NEH region," Indian Journal of Soil Conservation, vol. 26, no. 3, pp. 234–238, 1998.
13. N. P. Singh, O. P. Singh, and N. S. Jamir, Sustainable Agriculture Development Strategy for North Eastern Hill Region of India, Mittal, New Delhi, India, 1996, Edited by Shukla S. P. and Sharma N.
14. B. P. Bhatt, "Agroforestry for sustainable mountain development in N.E.H. region," inCentral Himalaya Environment and Development: Potentials, Actions and Challenges, M. S. S. Rawat, Ed., pp. 206–223, Transmedia Publisher, Uttaranchal, India, 2003.
15. Umashankar, "Indigenous agroforestry tree species for conservation and rural livelihood," in Agroforestry in North East India: Opportunities and Challenges, B. P. Bhatt and K. M. Bujarbaruah, Eds., pp. 149–174, ICAR Research Complex for NEH Region, Umiam, Meghalaya, 2005.
16. R. Saha, J. M. S. Tomar, and P. K. Ghosh, "Evaluation and selection of multipurpose tree for improving soil hydro-physical behaviour under hilly eco-system of north east India,"Agroforestry Systems, vol. 69, no. 3, pp. 239–247, 2007.

17. B. P. Singh and S. K. Dhyani, "Significance of jackfruit in restoration of soil fertility," Annual Report, ICAR Research Complex, Umiam, Meghalaya, 1995.
18. B. Majumdar, K. Kumar, M. S. Venkatesh, Patiram, and Bhatt B. P., "Effect of different agroforestry systems on soil properties in acid Alfisols of Meghalaya," Journal Hill Research, vol. 17, no. 1, pp. 1–5, 2004.
19. B. Majumdar, M. S. Venkatesh, K. K. Satapathy, K. Kumar, and Patiram, "Effect of alternative farming systems to shifting cultivation on soil fertility," Indian Journal of Agricultural Sciences, vol. 72, no. 2, pp. 122–124, 2002.
20. R. Saha and V. K. Mishra, "Long-term effect of various land use systems on physical properties of silty clay loam soil of N-E hills," Journal of the Indian Society of Soil Science, vol. 55, no. 2, pp. 112–118, 2007.
21. K. K. Satapathy, "Runoff production on hill slopes under different land use systems," inAgroforestry in North East India: Opportunities and Challenges, B. P. Bhatt and K. M. Bujarbaruah, Eds., pp. 451–459, ICAR Research Complex for NEH Region, Umiam, Meghalaya, 2005.
22. R. Saha, V. K. Mishra, and J. M. S. Tomar, "Effect of agroforestry systems on erodibility and hydraulic properties of Alfisols in eastern Himalayan region," Indian Journal of Soil Conservation, vol. 33, pp. 251–253, 2005.
23. R. Saha, P. K. Ghosh, V. K. Mishra, B. Majumdar, and J. M. S. Tomar, "Can agroforestry be a resource conservation tool to maintain soil health in the fragile ecosystem of north-east India?" Outlook on Agriculture, vol. 39, no. 3, pp. 191–196, 2010.
24. M. Datta and N. P. Singh, "Growth characteristics of multipurpose tree species, crop productivity and soil properties in agroforestry systems under subtropical humid climate in India," Journal of Forestry Research, vol. 18, no. 4, pp. 261–270, 2007.
25. S. L. Swamy, S. Puri, and A. K. Singh, "Growth, biomass, carbon storage and nutrient distribution in Gmelina arborea Roxb. Stands on red lateritic soils in central India,"Bioresource Technology, vol. 90, no. 2, pp. 109–126, 2003.

26. P. K. Ghosh, A. Das, R. Saha et al., "Conservation agriculture towards achieving food security in North East India," Current Science, vol. 99, no. 7, pp. 915–921, 2010.
27. R. Saha and V. K. Mishra, "Effect of organic residue management on soil hydro-physical characteristics and rice yield in eastern Himalayan region, India," Journal of Sustainable Agriculture, vol. 33, no. 2, pp. 161–176, 2009.
28. V. K. Mishra, R. Saha, and K. M. Bujarbaruah, "Zero tillage technique for transplanted rice in high rainfall eco-system," Scientific leaflet., ICAR Research Complex for NEH Region, Umiam, Meghlaya, 2005.
29. R. Saha, V. K. Mishra, B. Majumdar, K. Laxminarayana, and P. K. Ghosh, "Effect of integrated nutrient management on soil physical properties and crop productivity under a Maize (Zea mays)-mustard (Brassica campestris) cropping sequence in acidic soils of Northeast India," Communications in Soil Science and Plant Analysis, vol. 41, no. 18, pp. 2187–2200, 2010.
30. P. K. Ghosh, R. Saha, J. J. Gupta et al., "Long-term effect of pastures on soil quality in acid soil of north-east India," Australian Journal of Soil Research, vol. 47, no. 4, pp. 372–379, 2009.
31. K. Laxminarayana, B. P. Bhatt, and R. Tulsi, "Soil fertility buildup through hedgerow intercropping in integrated farming system: a case study," in Agroforestry in North East India: Opportunities and Challenges, B. P. Bhatt and K. M. Bujarbaruah, Eds., pp. 495–506, ICAR Research Complex for NEH Region, Umiam, Meghalaya, 2005.

Chapter 2

Role of Pigeonpea Cultivation on Soil Fertility and Farming System Sustainability in Ghana

S. Adjei-Nsiah

Forest and Horticultural Crops Research Centre, Kade, Institute of Agricultural

ABSTRACT

The productivity of the smallholder farming system in Ghana is under threat due to soil fertility decline. Mineral fertilizer is sparingly being used by smallholder farmers because of prohibitive cost. Grain legumes such as pigeonpea can play a complementary or alternative role as a source of organic fertilizer due to its ability to enhance soil fertility. Despite its importance, the potential of pigeonpea as a soil fertility improvement crop has not been exploited to any appreciable extent and the amount of land cultivated to pigeonpea in Ghana is very negligible. This paper synthesizes recent studies that have been carried out on pigeonpea in Ghana and discusses the role of pigeonpea cultivation in soil fertility management and its implication for farming system sustainability. The paper shows that recent field studies conducted in both the semi-deciduous forest and the forest/savanna transitional agro-ecological zones of Ghana indicate that pigeonpea/maize rotations can increase maize yield by 75–200%. Barrier to widespread adoption of pigeonpea include land tenure, market, and accessibility to early maturing and high yielding varieties. The paper concludes among other things that in order to promote the cultivation of pigeonpea in Ghana, there is the need to introduce varieties that combine early maturity with high yields and other desirable traits based on farmers preferences.

INTRODUCTION

Agricultural productivity in the smallholder farming systems in Ghana is under threat due to declining soil fertility. In the past, smallholder farmers in Ghana relied on the extended bush fallow system for maintaining the productivity of their farmlands [1]. This system allowed restoration of phosphorus (P) and nitrogen (N), the most limiting nutrients. However, over the years, the population growth-induced scarcity of suitable farmland has led to the shortening of the fallow period making it difficult to manage soil fertility in smallholder farming systems. The problem is compounded by the increasing cost of inputs at the farm level due to structural adjustment programmes that have removed subsidies and increased supply costs due to the deterioration conditions of rural infrastructure [2]. For instance, in 2002, whereas a metric tonne of urea cost about US$90 FOB (free on board) in Europe [3], the same quantity cost a Ghanaian farmer about US$308 at the farm level [4].

Most farmers, especially the smallholder farmers, do not have access to formal credit and therefore cannot afford to buy mineral fertilizers even when it has been demonstrated to be profitable [5]. Furthermore, due to the unpredictability of rainfall in some areas, farmers consider fertilizer application as risky.

In smallholder farming systems in Ghana, legumes can play a complementary or alternative role as source of organic fertilizer. Research in many parts of Sub-Saharan Africa including Ghana has shown that legumes have the potential to sustain soil fertility in smallholder farming systems [6–8]. Pigeonpea (Cajanus cajan (L.) Millsp.) has been found to have a great potential in this respect because of its ability to recycle nutrients and tolerate wide environmental conditions and low soil fertility [9]. Pigeonpea is nutritionally well balanced and is an excellent source of proteins (20–30%) [7]. In addition to proteins, pigeonpea provides carbohydrates and high levels of vitamins A and C.

Despite the importance of pigeonpea in terms of nutrition and soil fertility improvement, the crop has not been promoted to any appreciable extent by the national agricultural research and extension system in Ghana. In the forest/savannah transitional zone where the crop is widely cultivated, pigeonpea is considered as women's crop and therefore minor due to men's dominion over women in household decision-making process. Furthermore the overemphasis on maize in the current production systems as the major food security and cash crop in the forest/savanna transitional zone has relegated pigeonpea in particular and legumes in general as a minor crop.

However, recent studies in the forest/savannah transitional agroecological zone as well as the semideciduous forest zone of Ghana have demonstrated that when integrated in the cropping system as a form of rotation, pigeonpea has the potential of improving soil fertility. This paper discusses the role of pigeonpea cultivation in soil fertility management and its implication for farming system sustainability and food security.

MATERIAL AND METHODS

This paper is largely based on studies that have been conducted in the forest/savannah transitional agro-ecological zone of Ghana between 2002 and 2006 under the framework of Convergence of Science project [10] and between 2007 and 2010 under the IDRC funded Climate Change Adaption programme in Africa [11]. Results from the 2002–2006 studies have already been published [8, 9, 12], while results from 2007–2010 studies have not been published.

The Study Area and Population

The studies were conducted in the Wenchi Municipal (7° 27 and 8° 30 N, 1° and 2° 36 W) in the forest/savanna transitional agro-ecological zone of Ghana. The relief of Wenchi is gently undulating to flat. The soils, which are mainly Lixisols, are fragile with shallow top soils underlain with compact concretions and impermeable iron pans [8]. Temperatures are relatively high with a monthly mean of about 30°C. Rainfall is bimodal and starts in April and ends in November with a dry spell in August. The rainy season is followed by a long dry season from November to March. The annual rainfall is about 1,300 mm with about 107 rainy days. Wenchi municipal, which has a total population of 97,058 (2000 census), is ethnically diverse with about 20% of the population being migrants from the three northern regions of Ghana and the neighbouring Burkina Faso.

Research Approach and Methods

In 2002, we carried out a study under the Convergence of Sciences project to explore farmers' soil fertility management strategies and their relevant social context [8]. In order to ground the research in the needs of the farming communities, a diagnostic study was carried out in the study area between July, 2002 and July, 2003 using Participatory Rural Appraisal tools such as drawing of a community territory map (to identify the differences in soil fertility patterns), a transect walk (to reveal the diversity of the landscape), and analysis of soil fertility management strategies and group discussions. Group discussions with 10–40 farmers were held in the village centre and/or on farmers' fields.

In addition, two sets of individual interviews with farmers were conducted to collect qualitative and quantitative data. In the first interview, which involved 40 farmers, the selection of farmers was done through stratified sampling. A list of farmers in the community was obtained from the village committee secretary and every tenth name from the list was selected for individual interviewing. The second interview which involved 38 farmers was conducted later to look at the farming characteristics of the various subcommunities in the village using a wealth ranking exercise. For this interview, 6–10 persons were selected from each wealth category within each subcommunity. The individual interviews were semistructured in nature and served both to get more quantitative data on farm size, household composition, and the farming system and to obtain a better qualitative understanding of the soil fertility management strategies and their underlying rationale.

The diagnostic study was followed by farmer participatory on-farm experimentations with three farmer research groups established soon after the diagnostic study to evaluate the agronomic efficacy of the soil fertility management practices being used by the farmers [9]. Six cropping sequences: cassava cropping; pigeonpea cropping; mucuna/maize/mucuna rotation; cowpea/maize/cowpea rotation; maize/maize/maize; Imperata cylindrica fallow were evaluated on both farmer-managed and researcher-managed plots for their effects on soil fertility and yield of subsequent maize test crop. To deepen our understanding of soil fertility management, we carried out further exploration of diversity among the farmers according to gender, ethnicity, and wealth [12]. Farmers were selected from three communities in Wenchi according to ethnicity and gender for interview using semi structured questionnaires. We conducted two sets of interviews. For the first interview, the native households were categorized into male-headed and female-headed households. Subsequently, a stratified sample was selected consisting of 20 males from male-headed households, 20 females from male-headed households, and 20 females from female-headed households. In the case of the migrants, every farmer in the community was interviewed because of the small size of their population. As migrant women do not have their own farming enterprises, only males were interviewed. In the second interview, the farmers were selected through a wealth ranking exercise. Fifteen farmers were selected from each of three wealth categories for interviewing. In addition, focus group discussions were held with chiefs, community leaders, family heads, and opinion leaders about land tenure systems in Wenchi.

In 2008, under the IDRC, funded climate change adaptation in Africa project, farmers evaluated three early maturing and three late maturing pigeonpea varieties obtained from the International Crop Research Institute of the Semi-Arid Tropics (ICRISAT) in India, in a community in Wenchi called Asuoano.

RESULTS AND DISCUSSION

Farmers Views on Soil Fertility Decline and Their Causes

Farmers used several terms to express the fertility status of their soils although the terminology differed from one ethnic group to another. For instance, among the Akans and the Walas, when the soil loses its fertility, the farmers would say the soil is "tired" and therefore must be allowed to "rest" under bush fallow to regain its lost "energy." Farmers used various indicators such as colour, water holding capacity, and soil texture to assess soil fertility. A black soil was considered fertile while gravelly and sandy soils were considered less fertile. Other indicators used by farmers to express soil fertility included presence of earthworm casts which the farmers call "earth worm faeces," growth of crops, decline in crop yield and proliferation of certain plant species and weeds. For instance, the presence of weeds like chromolaena odorata and elephant grass (Pennisetum purpureum) indicated a fertile soil while the presence of weeds like spear grass (Imperata cylindrica) indicated a less fertile soil.

Farmers pointed out five major factors as being responsible for soil fertility decline in the area as follows: (1) increasing population as a result of migration of people into the area leading to smaller farms, which has resulted in the continuous cropping of the same piece of land over a long period; (2) annual bushfires which destroy vegetation and the population of earthworms that contribute to soil fertility improvement; (3) monocropping of maize without rotating it with any other crop; (4) the rapid increase in the monetary value of land; (5) felling of trees which exposes the land to the direct action of the sun.

Commonly Grown Crops in the Area

Table 1 shows the major food crops and grain legumes grown in the area. In terms of magnitude, maize was the most important food crop grown in Asuano. Other food crops included cassava, yams, cocoyam, and plantain. Legumes also played important role in the cropping systems in the area. Farmers listed about seven different types of legumes they cultivated (Table 1). In terms of proportion of land allocated to legumes, farmers ranked cowpea as the most important legume, followed by groundnut and pigeonpea. However, the type of legumes cultivated by farmers differed among the natives and the migrants. While the native farmers preferred pigeonpea, the migrant farmers preferred cowpea and groundnut. This was related to dynamics around land tenure. In Wenchi area, migrants do not own land but access land for farming through share cropping and land renting arrangements and hence preferred short duration legumes like cowpea and groundnut. Long duration legume such as pigeonpea was therefore not a preferred option for migrant farmers. Moreover, the market for pigeonpea is not well developed as it is a local food crop. It must be noted that the sum of the percentage number of people cultivating the various food crops added up to more than 100% because some farmers cultivated more than one crop since smallholder farmers in Ghana in general and Wenchi in particular practise mixed cropping. Yam was planted on mounds to increase the volume of soil for the development of tubers. Maize, cocoyam, and cassava were usually planted on the lower portions of the yam mounds while pigeonpea was planted in the rows between the yam mounds in a sequential order for efficient utilization of nutrients. Yam was usually planted first followed by maize, cocoyam, cassava, and lastly pigeonpea. The legumes mainly cowpea, groundnuts, and pigeonpea were grown to serve several purposes. These included (1) household food security; (2) household income; (3) soil fertility improvement through build up of soil organic matter and release of nutrients, particularly N for subsequent crops. Where crop rotation involving these legumes and maize was practised, crop succession was planned such that maize was planted after harvesting the legume to take advantage of the nitrogen fixed by the legume. In this way, it helped to either reduce or avoid the use of nitrogen fertilizers which may increase the cost of production.

Table 1: The most important food crops and proportion of land allocated to leguminous crops by native and migrant farmers in Asuoano in 2002

	Natives	Migrants
	%	
Food crop		
Maize	100	100
Cassava	100	88
Yam	100	94
Cocoyam	91	19
Plantain	32	0
Leguminous crop		
Pigeonpea	50	0
Groundnut	20	35
Cowpea	10	50
White kidney bean	10	0
Bambara groundnut	5	15
Mucuna	2	0
Others	3	0

Source: [8].

Role of Pigeonpea in the Cropping Systems in Wenchi

Among the legumes cultivated by the natives, pigeonpea was grown on a larger scale in comparison with other legumes because of its ability to regenerate soil fertility, its low production cost, its tolerance to pests and diseases, its cash income, and its food value. It was generally grown on less fertile land and land with problematic weeds such as spear grass (Imperata cylindrica).

After cropping a piece of land to crops like maize, cassava, and yam for about three to four years, farmers intercropped their food crops with pigeonpea during the last cropping year of the cycle. After harvesting the component crops, farmers allowed the land to remain under pigeonpea for 18–24 months after which the pigeonpea plants were cut down, burnt and the land cropped to maize or yams. The pigeonpea canopy was perceived to protect the soil from the direct action of the sun and therefore prevents the soil from becoming hardened. According to the farmers, pigeonpea forms a canopy after one year and shades out obnoxious weeds by suppressing their growth. The farmers also explained that the leaf litter covers the soil, reduces soil erosion, improves infiltration, prevents heating of the soil, and enhances earthworm activity. Crops grown on the land after pigeonpea, and especially maize, were perceived by the farmers to look greener, grow faster, and yield more. Rotations involving pigeonpea was often regarded by farmers as a long-term soil fertility management strategy. Despites its importance, migrant farmers preferred to use mainly short-term strategies such as mounding and the planting of short duration leguminous crops such as cowpea and groundnut for maintaining soil fertility because of the dynamics surrounding land tenure.

Agronomic Potential of Pigeonpea

Studies in Wenchi and Kade in the forest/savanna transitional and semideciduous forest zones of Ghana, respectively, indicate the potential of pigeonpea in improving soil fertility and farm profitability (Tables 2 and 3). In the transitional zone, rotating pigeonpea with maize resulted in about 75% increase in maize yield over the continuous maize control. In the semi-deciduous forest zone of Ghana, including pigeonpea in the cropping system also resulted in about 100–200% increase in maize grain yield over continuous maize. In the transitional zone, pigeon pea yielded about 25.5 tons of shoot biomass within 16 months of crop growth. High amounts of leaf litters of up to about 2 t ha^{-1} in one season could be considered a potential source of nutrients for subsequent crop as confirmed by a 75% increase in maize yield in the transitional zone and up to 200% increase in maize yield in the semi-deciduous forest zone of Ghana.

Table 2: Effect of crop sequence and N rate on (a) maize grain yield (kg ha^{-1}) and (b) weed biomass (kg ha^{-1}) associated with the maize crop at 8 weeks after planting on researcher-managed plots

(a)

Crop sequence	N rate (kg ha^{-1})		Mean
	O	60	
Speargrass fallow	1050	2848	1949
Cassava	3002	2738	2870
Pigeonpea	2422	2972	2697
Cowpea-maize-cowpea	1670	2328	1999
Mucuna-maize-mucuna	2970	4195	3582
Maize-maize-maize	1380	2128	1754
Mean	2082	2868	

SED: crop sequence (CS) = 318.4; N rate (NR) = 115.3; CS × NR = 375.8. P < F: CS = 0.001; NR = 0.001; CS × NR = 0.01.

Source [9].

(b)

Crop sequence	N rate (kg ha^{-1})		Mean
	O	60	
Speargrass fallow	585	790	686
Cassava	270	300	285
Pigeonpea	390	500	445
Cowpea-maize-cowpea	325	395	360
Mucuna-maize-mucuna	300	345	323
Maize-maize-maize	240	430	335
Mean	351	460	

SED: crop sequence (CS) = 65.9; N rate (NR) = 52.4; CS × NR = 112.1.

P < F: CS = 0.001; NR = 0.05; CS × NR = NS.

Source [9].

Table 3: Effect of crop sequence on maize grain yield at 12% moisture content

Crop sequence	Maize grain yield (t ha^{-1})	
	2007/2008	2008/2009
Pigeonpea	7.0	2.3
Ex-Subi	4.6	1.9
Boakentemma	3.5	1.4
Cowpea-Maize-Cowpea	3.0	1.5
Groundnut-Maize-Groundnut	2.7	1.6
Maize-Maize-Maize	2.3	1.3
LSD at 5%	1.4	0.58

Source [13].

Higher maize yield under legume-maize rotation is expected since legumes are known to fix nitrogen, thereby improving the soil nitrogen economy and enhancing the growth of subsequent crop [6, 14, 15] The higher maize grain yield associated with pigeonpea could largely be attributed to the large amounts of recycled N incorporated into the soil just before planting the maize test crops. The vigorous root system of pigeonpea has the capacity to explore large soil volume and recycles nutrients from deeper soil profiles. The faster decomposition of the biomass of pigeon pea and N release compared with that of maize stover was better synchronized with maize demand than the slower release of N by the poorer quality materials like maize stover. The N release patterns of organic residues of differential quality have been discussed extensively by [16, 17]. According to [16, 17], poor quality organic residues, with high C : N ratio, provide abundant supply of C for microbial growth leading to immobilization of soil N in the microbial biomass. Residues of high quality organic inputs on the other hand decompose quickly and may release about 70% of the N within a season under tropical conditions [18].

Besides the beneficial effect of pigeonpea on subsequent maize crops as a result of the high N recycling properties of the pigeonpea, the role of other nutrients cannot be ruled out. In some studies, increased soil P availability under pigeonpea was attributed to the efficient solubilisation and uptake of P from bound sources (e.g., Fe-P) by root exudates [19]. Benefit/cost analysis of different cropping sequences involving three grain legumes and cassava over four cropping seasons indicated that pigeonpea/maize rotations had 108% return on investment compared with 31% by continuous maize when fertilizer was not applied to the maize crop (Table 4).

Table 4: Estimated costs of production, gross revenue, and returns on investment of (a) various crop sequences (b) maize grown after the sequences with N application to the maize and (c) maize grown after the sequences without N application to the maize

Crop sequence	Economic yield (kg ha^{-1})	Total revenue (US$) ha^{-1}	Cost of production (US$) ha^{-1}			Total cost	Net revenue	Return on investment
			Land	Input	Labour			
(a) Crops in the sequence								
[1]Cassava	31,000	2545.1	41.7	41.7	635.0	718.4	1826.7	254
[2]Pigeonpea	1,870	623.3	41.7	8.3	221.5	271.5	351.8	130
[3]Mucuna-maize-mucuna	2,016	365.1	41.7	41.7	247.4	330.8	34.3	10
[4]Cowpea-maize-cowpea	2,536* (1,230)	1079.0	41.7	106,1	475.1	622.9	456.1	73
[5]Maize-maize-maize	3,287	595.2	41.7	36.1	386.1	463.8	456.1	28
[6]Speargrass fallow	0	0	41.7	0	0	41.7	−41.7	−100
(b) Maize after crop sequence with N application								
CS 1	2,738	495.9	13.9	104.2	190.2	308.3	187.6	61

CS 2	2,974	538.5	13.9	104.2	196.5	314.6	223.9	71
CS 3	4,194	759.4	13.9	104.2	245.9	364.0	395.4	108
CS 4	2,331	422.1	13.9	104.2	177.0	295.1	127.0	43
CS 5	2,126	385.0	13.9	104.2	175.4	293.5	91.4	31
CS 6	2,848	515.7	13.9	104.2	224.4	342.5	173.3	51
(c) Maize after crop sequence without N application								
CS 1	3,000	543.2	13.9	13.9	175.6	203.4	339.8	167
CS 2	2,423	438.8	13.9	13.9	165.5	193.3	245.5	127
CS 3	2,961	537.7	13.9	13.9	209.7	237.5	300.2	126
CS 4	1,772	302.8	13.9	13.9	155.2	183.0	119.8	66
CS 5	1,380	249.9	13.9	13.9	153.0	180.7	69.1	38
CS 6	1,048	189.8	13.9	13.9	173.7	200.9	−11.1	−6

[1]US$82.1 t−1.

[2]US$333.3 t−1.

[3]US$181.1 t−1 for maize.

[4]US$337.5 t−1 for cowpea and US$181.1 t−1 for maize.

[5]US$181.1 t−1 for maize.

[6]US$0.

*Yield of maize (figure in brackets).

CS 1: cassava; CS 2: Pigeonpea; CS 3: Mucuna-maize-mucuna; CS 4: Cowpea-maize-cowpea; CS 5: Maize-maize-maize; CS 6: Speargrass fallow.

Source: [9].

Challenges in the Cultivation of Pigeonpea

Barrier to widespread adoption of pigeonpea include land tenure, market, and availability of early maturing varieties. In the transitional zone of Ghana where the crop is widely cultivated by farmers, migrant farmers prefer cowpea cultivation to pigeonpea cultivation although they acknowledge the superiority of the latter over the former in terms of soil fertility management and the yield of subsequent maize crop (Table 5). According to the migrants, the market for pigeonpea is not always readily available and return on investment is too slow. In addition, they were afraid that when they invest in the soil, they would not be allowed to reap the full benefit. Among the native farmers, female and male farmers differed in the cultivation of legumes such as pigeonpea and cowpea. Women farmers preferred pigeonpea over cowpea due to its role in food security as well as its low labour requirement [12]. Male farmers preferred cowpea over pigeonpea because the market for cowpea is developed than that of pigeonpea.

Table 5: Preferential ranking of different soil fertility management practices by native and migrant farmers in Wenchi

Management practice	Ranking order*							
	Natives				Migrants			
	Asuoanoa *N* = 1 0	Beposob *N* = 5	Drobosoc *N* = 7	Average	Asuoanoc *N* = 6	Beposod *N* = 6	Drobosoe *N* = 5	Average
(a) Ranking by natives and migrants								
Cassava	1	1	1	1	2	2	1	1.7
Pigeonpea	2	5	2	3	4	4	4	4
Mucuna/Maize/Mucuna	7	6	4	5.7	5	6	6	5.6
Groundnut/Maize/Groundnut	4	3	3	3.3	3	3	3	3
Cowpea/Maize/Cowpea	3	2	5	3.3	1	1	2	1.3
Maize/Maize/Maize	8	7	6	7	7	7	7	7
Cowpea/Cowpea/Cowpea	5	4	7	5.3	6	5	5	5.3
Bush fallow	6	8	8	7.3	8	8	8	8
(b) Ranking by female and male Bonos	Females N = 13	Males N = 10						
Cassava	1	1						
Pigeonpea	2	3						
Mucuna/Maize/Mucuna	5	7						
Groundnut/Maize/Groundnut	3	4						
Cowpea/Maize/Cowpea	4	2						
Maize/Maize/Maize	8	8						
Cowpea/Cowpea/Cowpea	7	5						
Bush fallow	6	6						

[a]Consisted of 6 males and 4 females; [b]Consisted of 4 males and 1 female; [c]Consisted of 6 females and 1male; [d]Dagarbas; [e]Walas.

*Each treatment was compared directly against the other until they were ranked from the highest to the lowest with 1 being the highest ranking and 8 being the lowest.

Source: [9].

Access of farmers to early maturing varieties is also another barrier to widespread adoption of the pigeonpea. Although the late maturing indeterminate pigeonpea varieties have the higher potential to improve soil fertility than the early maturing varieties due to greater biomass production, farmers prefer the early maturing determinate varieties. Criteria farmers used for selecting varieties for planting included maturity, plant height (for ease of harvesting), tolerant to insects, plant architecture (prefer varieties with smaller plant canopy for ease of intercropping with yams), and seed size (bigger seed size is preferred). Amount of biomass and/litter falls and ability to improve soil fertility were not mentioned as criteria although farmers claimed that soil fertility was one of their major production challenges and that pigeonpea was one of the crops they used to improve the fertility of their soils. Although late maturing indeterminate pigeonpea varieties with woody stems are a potential source of fuel wood, farmers did not also include this in their selection criteria. This may be due to the fact that accessibility to fuel wood was not a major constraint in Wenchi.

During focus group discussions in the field, the farmers noted that the late maturing pigeonpea varieties had produced a lot of biomass and had higher litterfalls compared with the early maturing varieties. The farmers observed and commented on the greater litterfall from the late maturing varieties. They further stated that the greater production of biomass and the higher litterfall of the late maturing varieties would lead to better soil fertility improvement compared with the early maturing pigeonpea varieties. Despite their recognition of the soil fertility benefits of the late maturing varieties in improving soil fertility, the farmers did not include this as an important criterion for selecting which pigeonpea varieties to grow (Table 6). Mapfumo et al. [20] made a similar observation in Zimbabwe and suggested that pigeonpea is less likely to be adopted merely on the basis of its capacity to improve soil fertility. Farmers were thus interested in pigeonpea as a food security crop; soil fertility benefits were secondly. Farmers therefore opted for short duration/early maturing varieties which they could easily crop twice in a year. Thus the varieties ICPL 88039 and ICPL 88034 were the most preferred varieties because of their early maturity and high yielding. Although the variety ICPL 87091 was also preferred for their larger grain size, this variety was very susceptible to pod borers.

Table 6: (a) Farmers' preference ranking and (b) farmers' criteria for selection of pigeonpea varieties for planting

(a) Variety	Growth and yield characteristics					Ranking and reasons for the rank	
	Flowering pattern	Maturity period (days)	100 grains weight (g)	Yield (tons ha^{-1})	Seed colour	Ranking	Reasons for the rank
ICPL 88039	Determinate	125	10.1	1.6	Brownish red	1	Early maturing, short plant height, could be planted twice in a year
ICPL 88034	Determinate	125	10.2	1.8	Brownish red	2	Early maturing, short plant height, could be planted twice in a year
ICPL 87091	Determinate	125	12.1	1.1	white	3	Early maturing, short plant height, could be planted twice in a year
Maruti	Indeterminate	190	10.0	NA	Brownish red	4	Medium maturity, high litter fall
Lakshmi	Indeterminate	196	10.2	NA	Brownish red	4	High litter fall
ICPL 96053	Indeterminate	195	10.1	NA	Brownish red	4	High litter fall
(b) Criteria	Ranking						
Maturity	1						
Number of plantings in a year	2						
Easy of harvesting (plant height)	3						
Easy of intercropping with yams	4						
Seed size	5						

Implication of Large Scale Cultivation for System Sustainability

In addition to food uses, pigeonpea has outstanding soil improvement and conservation properties. The growth habit facilitates soil protection, as the canopy continues to expand during the dry season after the component crops in the mixed cropping have been harvested. Living and senesces pigeonpea leaves may offer protective cover for the soil during dry season to prevent the soil from drying out and reduce soil erosion and enhance rain percolation during the rainy season. In farming systems with minimal application of external inputs, management of organic resources plays a major role in maintaining both nutrient availability and soil organic matter [21]. In a cereal-based farming system as that found in the forest/savanna transitional agro-ecological zone of Ghana, where external input use is minimal, most recycling of N and P occurs through pigeonpea litterfall and green leafy biomass of pigeonpea incorporated into the soil after pigeonpea harvest [9]. Pigeonpea litterfall and green leafy biomass of pigeonpea are important sources of easily mineralizable N due to their high nitrogen (1.9 and 2.2% for litterfall and green leafy biomass, resp.) leading to high decomposition rates. Thus, rotation involving pigeonpea within smallholder agriculture has the potential of maintaining a reasonable supply of N and P to cereal crops, particularly maize considering the minimal use of external inputs in a maize-based farming system.

CONCLUSIONS

The paper shows the potential of pigeonpea in improving soil fertility in the smallholder farming systems particularly in a predominantly maize-based farming system as that found in Wenchi, Ghana. We have shown that the potential of pigeonpea in sustaining the productivity of predominantly maize-based farming system is due to its nutrient recycling properties as well as to its role in food security and flexibility in external input use and labour requirement. Soil fertility was apparently not a major criterion during varietal ranking by farmers suggesting that pigeonpea is less likely to be adopted merely on the basis of its capacity to improve soil fertility. Even when there was no strong market demand for pigeonpea, female farmers still integrated pigeonpea in their rotational system. Pigeonpea is largely a women's crop and therefore the attitude by men as major household decision makers has probably undermined its adoption in a predominantly maize based cropping system in Wenchi. Empowering women to make decision with respect to allocation of production resources may enhance a greater adoption of legumes. As more farmers resort to putting their land under pigeonpea instead of bush fallowing, pigeonpea cultivation could serve as an entry point for farming system sustainability. There is, however, the need to (i) introduce short duration pigeonpea varieties that will fit well in the predominantly mixed cropping systems of smallholder agriculture; (ii) evaluate the nutrient recycling capacities of different pigeonpea genotypes; (iii) develop crop rotation/sequencing and soil management options that can improve and/or sustain the productivity of pigeonpea through integrated soil fertility management (ISFM); (iv) design a range of social arrangements that will encourage investment in soil fertility through integration of pigeonpea in the farming system in a heterogeneous farming community like Wenchi, (v) develop use for pigeonpea in new cooking options and/or value added products.

ACKNOWLEDGMENTS

The author is grateful to the two anonymous reviewers who reviewed an earlier version of this paper. Financial support towards this work provided by the Interdisciplinary Research and Education Fund (INREF) of Wageningen University and Research Centre, and The Netherlands and the Dutch Ministry for International Cooperation (DGIS) is gratefully acknowledged. The author also thanks the International Development Centre for funding the climate change adaption in Africa project through the University of Zimbabwe.

REFERENCES

1. P. H. Nye and D. Stephen, "Soil fertility," in Agriculture and Landuse in Ghana, J. B. Wills, Ed., pp. 127–143, Oxford University Press, London, UK, 1962.
2. H. Gerner, E. O. Asante, E. Owusu Benoah, and K. Marfo, Ghana Fertilizer Privatization Scheme. Private Sector Role and Public

Sector Responsibility in Meeting Needs of Farmers, IFDC-Africa, Lome, Togo, 1995.

3. P. A. Sanchez, "Soil fertility and hunger in Africa," Science, vol. 295, no. 5562, pp. 2019–2020, 2002. View at Publisher · View at Google Scholar · View at Scopus
4. ISSER, The state of Ghana economy in 2004, University of Ghana, Legon, Ghana, 2005.
5. H. B. Obeng, K. G. Erbyn, and E. O. Asante, "Fertilizer requirements and Use in Ghana," Consultancy Report, Tropical Agricultural Development Consultancy, Accra, Ghana.
6. K. E. Giller, Nitrogen Fixation in Tropical Cropping Systems, CAB International, Wallingford, UK, 2nd edition, 2001.
7. S. S. Snapp, R. B. Jones, E. M. Minja, J. Rusike, and S. N. Silim, "Pigeon Pea for africa: a versatile vegetable—and more," HortScience, vol. 38, no. 6, pp. 1073–1079, 2003.
8. S. Adjei-Nsiah, C. Leeuwis, K. E. Giller et al., "Land tenure and differential soil fertility management practices among native and migrant farmers in Wenchi, Ghana: implications for interdisciplinary action research," NJAS-Wageningen Journal of Life Sciences, vol. 52, no. 3-4, pp. 331–348, 2004.
9. S. Adjei-Nsiah, T. W. Kuyper, C. Leeuwis, M. K. Abekoe, and K. E. Giller, "Evaluating sustainable and profitable cropping sequences with cassava and four legume crops: effects on soil fertility and maize yields in the forest/savannah transitional agro-ecological zone of Ghana," Field Crops Research, vol. 103, no. 2, pp. 87–97, 2007.
10. D. Hounkonnou, D. K. Kossou, T. W. Kuyper et al., "Convergence of sciences: the management of agricultural research for small-scale farmers in Benin and Ghana," NJAS-Wageningen Journal of Life Sciences, vol. 53, no. 3-4, pp. 343–367, 2006.
11. P. Mapfumo, F. Matambanengwe, R. Chikowo et al., "Farmers› perceptions lead to experimentation?"LEISA Magazine, vol. 24, 2008.
12. S. Adjei-Nsiah, C. Leeuwis, O. Sakyi-Dawson, K. E. Giller, and T. W. Kuyper, "Exploring diversity among farmers for orienting inter-disciplinary action research on cropping system management in Wenchi, Ghana: the significance of time horizons," International Journal of Agricultural Sustainability, vol. 5, pp. 176–194, 2007.

13. S. Adjei-Nsiah, "Evaluating cropping sequences with cassava and three grain legume crops: effects on soil fertility and maize yields in the semi-deciduous forest zone of Ghana," Journal of Soil Science and Environmental Management, vol. 4, no. 1, 2012.
14. S. Adjei-Nsiah, Cropping systems, land tenure and social diversity in Wenchi, Ghana: implication for soil fertility management [Ph.D. thesis], Wageningen University, The Netherlands, 2006.
15. A. A. Yusuf, R. C. Abaidoo, E. N. O. Iwuafor, O. O. Olufajo, and N. Sanginga, "Rotation effects of grain legumes and fallow on maize yield, microbial biomass and chemical properties of an Alfisol in the Nigerian savanna," Agriculture, Ecosystems and Environment, vol. 129, no. 1–3, pp. 325–331, 2009.
16. S. Recous, D. Robin, D. Darwis, and B. Mary, "Soil inorganic N availability: effect on maize residue decomposition," Soil Biology and Biochemistry, vol. 27, no. 12, pp. 1529–1538, 1995.
17. W. D. Sakala, G. Cadisch, and K. E. Giller, "Interactions between residues of maize and pigeonpea and mineral N fertilizers during decomposition and N mineralization," Soil Biology and Biochemistry, vol. 32, no. 5, pp. 679–688, 2000.
18. K. E. Giller and G. Cadisch, "Future benefits from biological nitrogen fixation: an ecological approach to agriculture," Plant and Soil, vol. 174, no. 1-2, pp. 255–277, 1995.
19. N. Ae, J. Arihara, K. Okada, T. Yoshihara, and C. Johansen, "Phosphorus uptake by pigeon pea and its role in cropping systems of the Indian subcontinent," Science, vol. 248, no. 4954, pp. 477–480, 1990.
20. P. Mapfumo, B. M. Campbell, S. Mpepereki, and P. Mafongoya, "Legumes in soil fertility management: the case of pigeonpea in smallholder farming systems of Zimbabwe?" African Crop Science Journal, vol. 9, no. 4, pp. 629–644, 2001.
21. C. A. Palm, K. E. Giller, P. L. Mafongoya, and M. J. Swift, "Management of organic matter in the tropics: translating theory into practice," Nutrient Cycling in Agroecosystems, vol. 61, no. 1-2, pp. 63–75, 2001.

Chapter 3

N, P, and K Budgets and Changes in Selected Topsoil Nutrients over 10 Years in a Long-Term Experiment with Conventional and Organic Crop Rotations

Audun Korsaeth

Arable Crops Division, Norwegian Institute for Agricultural and Environmental Research, 2849 Kapp, Norway

ABSTRACT

This study presents soil system budgets of N, P and K in six contrasting cropping systems during 10 years of a long-term experiment in southeast Norway. The experiment included systems with arable cash-cropping and with mixed arable-dairy cropping (cash- and fodder crops), with organic and conventional management represented in both groups. All major nutrient inputs and outputs were measured or estimated. State of the art conventional cash-cropping appeared to be balanced in terms of N, whereas conventional mixed cropping had an N surplus. By contrast, less up to date conventional arable cash-cropping and all the organic systems showed indications of soil organic N depletion (negative N budgets). All the organic systems showed that mining of the soil P and K content occurs, whereas the conventional systems all had P and K surpluses. The results corresponded well with measured differences between systems in terms of ignition loss, P-AL, K-AL and K-HNO_3 measured in 2009. This study shows that a fertile soil may be exposed to substantial mining of N, P and K over many years before it is detectable by traditional analyses, and that field nutrient budgeting is a feasible, but data-demanding, approach to detect such misbalances at an early stage.

INTRODUCTION

In 1989, a large cropping system experiment, facilitated for measurements of runoff and leaching, was established at Apelsvoll in southeast Norway. Over the years, this experiment has provided data for many studies covering a range of different topics, including yields and yield quality (e.g., [1]), nutrient leaching and runoff losses (e.g., [2]), economic aspects (e.g., [3]), soil microbiology (e.g., [4]), soil physical and chemical properties (e.g., [5]), and the relation between food production and N losses [6]. Some major adjustments of the experimental design were made in 2000 [6]. In this overview, a synthesis of the results obtained after these changes are given for the major nutrient flows of N, P, and K, with focus on changes in topsoil nutrient pools, as affected by misbalances between nutrient inputs and outputs at field level. Numerous long-term experiments have shown that crop rotation and management affect soil fertility (e.g., [7–13]). However, considerable time is needed before identifiable changes in soil fertility emerge [14]. Nutrient budgets have been used widely in a range of farming systems to assess long-term sustainability (e.g., [15]), thus, supplementing soil measurements. In a discussion of uncertainties in nutrient budgets, Oenema et al. [16] distinguished between farm gate soil surface and soil system budgets. The latter accounts for nutrient inputs and outputs, recycling of nutrients within the system, nutrient loss pathways, and changes in soil nutrient pools. Soil system budgets were considered to possess the highest uncertainty of the three budgeting approaches, since nutrient losses via leaching, runoff, volatilization, and denitrification are classified as the most uncertain nutrient flows [17]. De Vries et al. [18], when estimating uncertainties in the soil system N budget of The Netherlands, reported that leaching to ground water and leaching to surface water had the highest relative uncertainty (coefficient of variation). Acquiring quality data on nutrient drainage and runoff may, therefore, considerably reduce the uncertainties of the soil system budget approach. The present study is aimed at comparing the effects of management (i.e., organic versus nonorganic) and type of production (i.e., arable cash cropping versus mixed dairy farming) on their long-term sustainability in terms of plant nutrition, by a combination of soil system nutrient budgeting and soil measurements. Results on drainage discharge and water-borne nutrient losses will be presented in more detail elsewhere [19].

MATERIAL AND METHODS

Experimental Site and Treatments

In 1989, a 3.2 ha large experiment with pipe-drained plots was established at Apelsvoll Research Centre in central southeast Norway (60°42′N, 10°51′E, altitude 250 m). The climate of the region is humid continental with a mean annual precipitation of 600 mm and a mean annual temperature of 3.6°, and 12.0°C in the growing season (May to September). On the experimental area, which slopes 2–8% northeast, deforestation was performed in 1935, whereafter it was used as pasture until 1975. During the following years, up to the establishment of the experiment in 1988, the field was cropped with a rotation including 10% root crops, 40% cereals, and 50% ley, using an average of 10 tonnes cattle slurry $ha^{-1} yr^{-1}$ plus regular amounts of inorganic fertilizer. The first year after draining the experimental site (1989), the area was cropped with barley (Hordeum distichum L.). The major soil group of the experimental area is classified as Endostagnic Cambisol [20], with dominantly loam and silty sand textures. More detailed soil characteristics have been presented by Riley and Eltun [21]

Table 1: Mean values of selected soil physical parameters in topsoil (0–30 cm) measured in 1988, prior to the start of the experiment (from [21])

System	Parameter number[1]									
	1	2	3	4	5	6	7	8	9	10
CA1	51	10	11	29	12	7	6	43	38	19
CA2	48	10	6	27	11	3	9	52	33	15
OA	49	7	7	30	12	4	6	46	35	19
CM	48	9	8	27	12	5	9	47	35	19
OM1	49	7	9	30	12	6	8	50	33	17
OM2	50	8	7	30	12	4	5	45	36	19
s.e.[2]	1.5	1.2	2.5	1.2	0.5	2.1	1.5	1.5	1.0	1.0
Mean	49	9	8	29	12	5	7	47	35	18

[1]Parameter number: (1) total porosity (%), (2) air capacity at pF2 (%), (3) air permeability at pF2 (μm^2), (4) total available water (pF 2–4.2) (%), (5) nonavailable water (pF > 4.2) (%), (6) hydraulic conductivity (cm/h), (7) gravel content (%), (8) sand content (%), (9) silt content (%), (10) clay content (%).

[2]Standard error of means

The experimental site comprises 12 blocks (30 × 60 m), separated by 7.5 m grass border zones (Figure 1). In each block, surface runoff is collected at the lower end and led to a sedimentation tank, and the blocks are separately drained with PVC pipes at a depth of 1 m with 7.5 m spacing. Surface runoff from the sedimentation tank and drainage water is transported in sealed plastic pipes to measuring stations equipped for discharge measurements (tipping buckets) and volume proportional sampling.

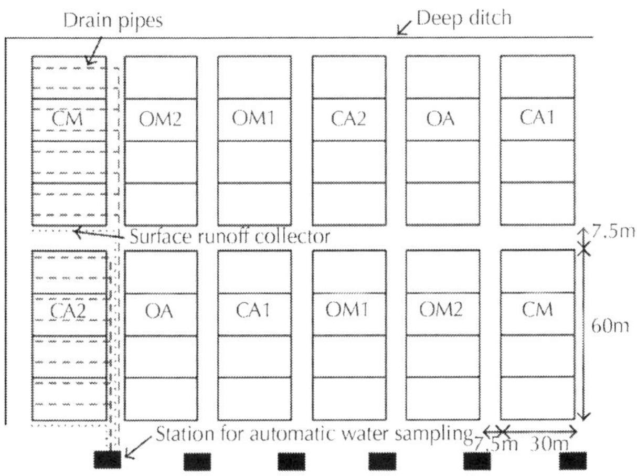

Figure 1: Layout of the cropping system experiment at Apelsvoll.

Using a randomized complete block design, six cropping systems, each with 2 replicates, were established on the twelve blocks in 1989. The first ten years (1989–1999) the experiment comprised three arable systems (conventional and integrated arable cropping without farmyard manure and organic cropping with some farmyard manure) and three mixed dairy systems (conventional, integrated, and organic production of both arable and forage crops, all with farmyard manure). Each block comprised eight 7.5 × 30m plots, on which all of the arable and/or fodder crops in the rotation were grown each year.Some major adjustments of the experimental design were made in 2000. The number of rotation plots was reduced from eight to four by merging pairs of neighbouring plots, thus, reducing rotation length from eight to four years, but still with each crop present every year. A new organic mixed dairy system was introduced instead of the integrated mixed dairy system, and some smaller changes were made in the management of the other systems. The six adjusted cropping systems are described briefly below (see Table 2 for details).

Table 2: Characteristics of the cropping systems at Apelsvoll, 2001–2010

Crop rotation	Fertilizer, kg ha^{-1}			Slurry, kg ha^{-1}			Plant protection	Soil tillage
	N	P	K	N	P	K		
Conventional arable (CA1)[1]								
Potatoes	119	51	183				Chemical, mechanical	Autumn ploughing and spring harrowing
Spring wheat	141	25	66				Chemical	
Spring oats	120	22	57				Chemical	
Spring barley	120	22	57				Chemical	
Conventional arable (CA2)								
Potatoes	108[7]	47	164				Chemical, mech.	Spring harrowing only[7]
Wheat + catch crop[3]	134[2, 4]	29	80				Chemical, mech.[5]	
Oats + catch crop[6]	116[2]	21	55				Chemical, mech.[5]	
Barley + catch crop[3]	116[2]	21	55				Chemical, mech.[5]	
Organic arable (OA)								
Barley[8]							Manual weeding	Spring ploughing and harrowing
Grass clover[9]							Manual w.	
S. wheat + catch crop							Manual w., mech.[5, 10]	
Oats + peas							Manual w., mech.[10]	
Conventional mixed dairy farming (CM)								
Barley[11]	50[2]	9	24	70	10	90	Chemical	Spring ploughing and harrowing
1st ley year	99[2]	16	75	57	8	73	—	
2nd ley year	114[2]	19	86	80	11	104	—	
S. wheat + catch crop	86[2, 4]	16	43	69	10	90	Chemical, mech.[5]	

Organic mixed dairy farming (OM1)								
Barley[11]				75	11	98	Manual w.	Spring ploughing and harrowing
1st ley year				24	2	31	Manual w.	
2nd ley year				48	6	63	Manual w.	
S. wheat + catch crop				83	12	107	Manual w., mech.[5, 10]	
Organic mixed dairy farming (OM2)								
Barley[11]				93	13	112	Manual w.	Spring ploughing and harrowing
1st ley year				63	8	80	Manual w.	
2nd ley year				98	13	122	Manual w.	
3rd ley year				75	10	93	Manual w.	

[1] Managed as was common for the region in 1985 (tillage and fertilization as in 1985, but for practical reasons, present-day inputs of seeds and chemical plant protection).

[2] Fertilizer level in spring adjusted according to regional recommendations, based on measurements of mineral N in topsoil in early spring. Average values are shown.

[3] Perennial ryegrass (Lolium perenne L), sown about one week after the cereals.

[4] Split application of fertilizer with about 75% given at sowing, and 0–60 kg N ha−1 applied at growth stage (GS) 49, according to measurements with the N tester [6].

[5] Weed harrowing performed when the cereals are at GS 11-12.

[6] Italian ryegrass (Lolium multiflorum Lam), sown about one week after the oats.

[7] Performed twice with a horizontally rotating harrow.

[8] With undersown grass-clover mixture. Seed mix: 80% Timothy (Phleum pratense L.), 10% red clover (Trifolium pratense L.), and 10% white clover (Trifolium repens L).

[9] Green manure, not harvested but mulched 3-4 times per season.

[10] Harrowed in autumn after harvest some years to reduce the weed pressure.

[11]With undersown grass-clover ley. Seed mix: 60% Timothy (Phleum pratense L.), 30% Meadow fescue (Festuca pratensis L.), and 10% red clover (Trifolium pratense L.).

CA1

Conventional arable cropping, managed as was common for the region in 1985 (tillage and fertilization as in 1985, but for practical reasons, present-day inputs of seeds and chemical plant protection). The year 1985 was selected, since the North-Sea Agreement (1987) used this year as the base for its planned 50% reduction in nutrient leaching to the North Sea within a 10-year period. Before this date, less attention was paid to nonpoint source losses of nutrients attributed to farming activities, and this cropping system is, thus, used as a reference.

CA2

Conventional arable cropping, using currently available knowledge in order to minimize the ratio of N lost by leaching and runoff to production. This optimisation involves the use of catch crops, split application of fertilizer, and reduced noninversion tillage.

OA

Organic arable cropping without cattle slurry, but with 25% of the area used for green manure (grass clover ley).

CM

Conventional mixed dairy farming, optimised similarly to CA2, but with spring ploughing, 50% of the area as grass clover ley and the use of slurry (amounts calculated from the theoretical number of cows sustained, see Section 2.3 for details).

OM1

Organic mixed dairy farming with 50% of the area as grass clover ley and slurry use (amounts calculated as in CM).

OM2

Organic mixed dairy farming with 75% of the area as grass clover ley and slurry use (amounts calculated as in CM).

The results in 2000 were partly influenced by the previous management. Therefore, this paper deals with the results from the decade May 2001–April 2011

Measurements

Within each plot, dry matter (DM) yields of grass, grain, straw (when removed), and potato tubers were measured in quadruplicate (subplot size 1.5 × 6 m). Straw was removed from all the cereal plots of CA1 and from the plots with barley undersown with grass clover ley in OA, CM, OM1, and OM2. Crude protein contents of the cereals were measured by near infrared reflectometry (INFRA 250, Technicon, USA). Potato tuber size distribution and quality parameters were determined according to standard procedures. The proportion of legumes in the grass clover ley was determined visually before harvest.

From 2006 onwards, plant samples of all harvested crops (0.2 g DM) were digested in a mixture of sulfuric acid and hydrogen peroxide and analysed for N and P colorimetrically with an autoanalyser (Skalar 5100, Actlabs, Canada) and for K by flame photometry (Corning 400, Sherwood Scientific Ltd., UK). Cattle slurry was sampled 1-2 weeks before application and analysed for total-N using the Kjeldahl method. Ammonium-N and nitrate-N were extracted with 2 M KCl and determined colorimetrically with an autoanalyser (Traacs, Bran and Luebbe, Germany). The water samples (drainage water and surface runoff) were analysed on a monthly basis for total N, ammonium-N, nitrate-N, total P, phosphate P, and total K and determined spectrophotometrically (DR2800 spectrophotometer, Hach Lange, Germany). Potassium was first included in May 2009. Soil samples have been taken every 3–5 years since 1989. Due to differences in sampling depths, sampling locations, and parameters analysed, only selected samples are comparable. In this study results are shown for topsoil samples (0–25 cm depth) taken in 1996, 1999, 2003, and 2009. The samples taken in 1999 and 2009 were analysed for ignition loss (at 550°C), whereas samples taken in 1996, 2003, and 2009 were used to quantify plant available P and K and acid soluble K. Plant available P (P-AL) and K (K-AL) was extracted by a mixture of acetic acid and ammonium lactate, according to Egnér et al. [22], whereas acid soluble K (K-HNO_3) was extracted by boiling in 1 M HNO_3. The P and K concentrations in the extracts were analysed by inductively coupled plasma (ICP) techniques (SPECTRO GENESIS, Analytical Instruments GmbH, Germany)

Calculations and Estimates

Dry and wet atmospheric N deposition was set to 2.7 and 7.2 kg N ha^{-1} yr^{-1}, respectively, and wet atmospheric K deposition was set to 2.1 kg ha^{-1} yr^{-1}, based on measurements at the nearest monitoring station, Hurdal [23], about 50 km south of Apelsvoll. Dry depositions of K and wet and dry P depositions were assumed to be negligible.

Symbiotic N fixation was estimated in accordance with Korsaeth and Eltun [2]. Nonsymbiotic N fixation was considered to be negligible. Nutrient inputs with seeds were estimated using measured N, P, and K contents of harvested grain and potatoes from the CON-A system. Literature values were selected for legumes and grasses.

The amounts of N in cattle slurry applied were calculated as the sum of N in harvested forage (grass clover ley) and feed concentrates (purchased and/or produced on the farm), minus the estimated N losses occurring from harvest until slurry application (forage losses, gaseous losses from the cow shed, and during slurry storage) and N exports via milk and livestock, as described in detail by Korsaeth [6]. The number of cows which each farming system could sustain was calculated from the average total available feed in the system during the previous three years (sliding mean) and the feed requirement for milk production, maintenance, activity and replacement. On the conventional farm, it was assumed that purchased cereal-based feed concentrates corresponded to 25% of the total fodder units available (feed concentrates plus forage grass). The organic systems were assumed to be completely self-sufficient, using on-farm produced barley as feed concentrates, for at most 20% of the total feed requirement. Barley not used as concentrates was assumed to be sold.

The yearly average cereal yields of OA were reduced by 25% to correct for the area used for green manure production. In 2007, the mixture of oat and peas in OA was replaced in the rotation by faba bean (Vicia fabaL.). These beans were totally damaged by chocolate spot (Ascochyta blight disease) and not harvested. All OA data from 2007 were, thus, excluded from further analyses.

Nutrient concentrations (N, P, and K) in harvested products (crops and straw) for the years 2001–2005 were set equal to the averages (separately for each cropping system) for the years 2006–2010. Removal of N, P, and K was calculated as the measured dry weight of products removed from the field multiplied by the estimated nutrient concentrations. Amounts of N, P and K in harvested grass were reduced by 10% to correct for likely losses under practical harvest.

Gaseous N-emissions (N_2O-N, -N, and NH_3-N) were estimated from the IPCC framework [24], which comprises estimates for both direct and indirect emissions. Nitrogen sources included in the direct estimates used here were mineral N fertilizer, applied N in slurry, and N in above-ground and below-ground crop residues. Net N mineralization associated with possible loss of SOM resulting from contrasting management was not considered. The volatilization of NH_3-N (and -N) is in the IPCC framework [24] related to the input of mineral fertilizer and organic N additions, not including crop residues. This implies that the mulched grass clover in OA would have zero emissions of NH_3-N using the IPCC-approach, which is very unrealistic (e.g., [25]). Hence, the volatilization of NH_3-N from this crop was calculated by means of a separate method [6].

Nutrient runoff occurring during each agrohydrological year, lasting from 1 May to 30 April, was attributed to the cropping season within that period. Calculations of N, P, and K transported via surface runoff and drainage water were based on measured nutrient concentrations and volumes of surface and drainage water. Organic N was calculated as the difference between total N and the sum of ammonium-N and nitrate-N. Potassium runoff occurring during the agrohydrological years 2001–2008 were set equal to the measured average K runoff for the agrohydrological years 2009/10 and 2010/11.

Nutrient soil system budgets were calculated separately for each system by considering all major flows of N, P, and K, respectively, with the above-ground crops representing the upper boundary and the drain pipes the lower boundary. A positive soil system budget, that is, where the inputs exceeded the outputs, was taken as an indication of nutrient accumulation, whereas a negative budget was taken as an indication of soil mining of the nutrient in question.

Statistics

Analyses of variance (ANOVA) were performed on yields and nutrient concentrations, using a split-plot model with cropping system as main plot and year as subplot. Grass-clover ley yields were analysed as the sum of two cuts, whereas differences in nutrient concentrations were analysed for each cut separately. Paired comparisons (LSD) were performed [26]. Comparisons of soil chemical properties measured on different occasions were conducted using the paired Student's t-test. In all tests, significance was assumed at P levels < 0.05. Mean data are presented with their standard errors

RESULTS

Yields

There were significant yield differences between cereal crops within each group of cropping systems (Table3). The conventional arable systems (CA1 and CA2) gave the largest overall cereal yields; the conventional mixed dairy system (CM) was intermediate, whereas the organic systems gave the lowest yields. The organic arable system had the lowest (area corrected) yields overall, achieving only 40 and 44%, respectively, of the barley and wheat harvested in the arable conventional systems. The mixture of oats and peas in OA compared slightly more favorably, but still with only 47% of the yield level obtained with monocropped oats in CA1 and CA2 (Table 3)

Table 3: Harvested yields of cereals and potatoes and their concentrations of N, P, and K

Crop	System				Nutrient concentrations (g 100 g DM^{-1})								
		Yield[1] (kg ha^{-1})			N			P			K		
		Mean	s.e.[2]	c[3]	Mean	s.e.	c	Mean	s.e.	c	Mean	s.e.	c
Barley	CA1	5450	152	a	2.05	0.061	a	0.40	0.025		0.39	0.008	ab
	CA2	5181	148	ab	1.99	0.054	ab	0.40	0.027		0.37	0.006	b
	OA	2148	125	d	1.68	0.034	c	0.41	0.015		0.39	0.009	ab
	CM	4865	134	b	1.85	0.034	bc	0.39	0.019		0.37	0.006	b
	OM1	3881	130	c	1.77	0.035	bc	0.40	0.019		0.40	0.003	ab
	OM2	4017	153	c	1.94	0.042	ab	0.42	0.020		0.41	0.006	a
	LSD	347			0.18			n.s.			0.02		
Oats	CA1	5866	260	a	2.07	0.050		0.38	0.020		0.37	0.006	
	CA2	5264	214	a	1.99	0.059		0.39	0.012		0.39	0.007	
	OA[4]	2618	297	b	2.30	0.155		0.43	0.012		0.59	0.012	
	LSD	944			n.s.			n.s.			n.s.		
Wheat	CA1	5696	217	a	2.25	0.047		0.40	0.023		0.40	0.002	
	CA2	5606	252	a	2.30	0.037		0.41	0.018		0.39	0.008	
	OA	2494	185	d	2.23	0.070		0.45	0.012		0.39	0.007	
	CM	4900	260	b	2.34	0.044		0.42	0.014		0.39	0.005	
	OM1	3719	187	c	2.21	0.078		0.42	0.013		0.39	0.009	
	LSD	574			n.s.			n.s.			n.s.		

Potatoes	CA1	9474	328		1.20	0.016		0.30	0.004		1.45	0.047	
	CA2	9535	349		1.22	0.038		0.28	0.001		1.42	0.058	
	LSD	n.s.			n.s.			n.s.			n.s.		

[1]Yields of cereals were corrected to a water content of 15%, whereas yields of potatoes are presented as DM.

[2]Standard error of means. Yield data were averaged for the years 2001–2010, whereas data on nutrient concentrations were averaged for the years 2006–2010.

[3]Pairwise comparisons of yields/nutrient concentrations using LSD at the 5% level, where yields/nutrient concentration of the same crop differing significantly between systems are denoted different letters. Nonsignificant comparisons are denoted n.s.

[4]In the organic system OA, oat was grown in a mixture with peas. Yields are given as the sum of the two crops, whereas nutrient concentrations are given as a weighted average (based on the DM weights of the two crops).

The total fresh weight yield of potatoes was 43 $\pm 1.3 \text{Mg ha}^{-1}$, and there was no significant difference in yield (Table 3), size distribution, or selected quality parameters between the two cropping systems with potatoes in the rotation (CA1 and CA2, data not shown). The DM content was 0.224 ± 0.005 kg DM kg fresh weight^{-1}, and about 93% (weight basis) of the potato tubers were saleable. The conventional mixed dairy system (CM) had significantly larger grass clover ley yields than the organic systems, both for the 1st and the 2nd ley years (Table 4). There was no significant yield difference between the organic systems. Their total production in the two first ley years was 86% of that obtained conventionally.

Table 4: Harvested ley yields, clover content and concentrations of N, P, and K.

Year	System Cut	Yield (kg DM ha^{-1})			Clover[1] (%)			Nutrient concentrations (g 100 g DM^{-1}) N			P			K		
		Mean	s.e.[2]	c[3]	Mean	s.e.	c[4]	Mean	s.e.	c[5]	Mean	s.e.	c	Mean	s.e.	c
Yr 1	CM	9630	335	a												
	Cut 1	4765	171		7	1.4	b	1.50	0.06		0.22	0.01		1.84	0.09	
	Cut 2	4865	243		10	2.4	B	1.48	0.09	B	0.28	0.02		1.98	0.14	
	OM1	8203	319	b												
	Cut 1	4092	218		27	2.6	a	1.55	0.09		0.22	0.01		1.69	0.06	
	Cut 2	4111	248		45	4.7	A	2.24	0.05	A	0.31	0.02		2.10	0.10	
	OM2	8342	278	b												
	Cut 1	4099	194		27	2.4	a	1.61	0.10		0.24	0.016		1.68	0.11	
	Cut 2	4243	228		40	5.0	A	2.16	0.16	A	0.32	0.01		2.02	0.08	
	LSD[6]	710			12			n.s.			n.s.			n.s.		
	LSD[7]				11			0.45			n.s.			n.s.		
Yr 2	CM	11021	304	a												
	Cut 1	6269	180		11	2.1	b	1.57	0.04		0.24	0.01		1.99	0.08	a
	Cut 2	4752	233		11	2.1	A	1.46	0.06	B	0.27	0.02		1.82	0.09	
	OM1	9542	264	b												
	Cut 1	5453	188		34	2.5	a	1.64	0.11		0.22	0.01		1.65	0.06	b
	Cut 2	4089	201		42	3.6	B	2.19	0.04	A	0.30	0.01		1.76	0.07	
	OM2	9586	312	b												

	Cut 1	5399	209		30	2.9	a	1.64	0.10		0.25	0.01		1.69	0.06	b
	Cut 2	4187	189		36	2.5	B	2.22	0.13	A	0.30	0.01		1.84	0.05	
	LSD[6]	892			10			n.s.			n.s.			0.25		
	LSD[7]				13			0.51			n.s.			n.s.		
Yr 3	OM2	8651														
	Cut 1	5499	169		19	2.3		1.52	0.05		0.24	0.01		1.60	0.08	
	Cut 2	3151	217		24	2.8		2.11	0.06		0.31	0.01		1.65	0.09	

[1]The proportion of legumes was determined visually before harvest.

[2]Standard error of means. Yield data were averaged for the years 2001–2010, whereas data on nutrient concentrations were averaged for the years 2006–2010.

[3]Pairwise comparisons of yields (sum of two cuts) using LSD at the 5% level, where yields of the same crop not differing significantly between systems are denoted the same letter.

[4]Pairwise comparisons of clover content separate for each cut, where clover contents of the first cut not differing significantly between systems are denoted the same letter, whereas clover contents of the second cut not differing significantly between systems are denoted the same capital letter.

[5]Pairwise comparisons of nutrient concentrations separate for each cut, where concentrations of the first cut not differing significantly between systems are denoted the same letter, whereas concentrations of the second cut not differing significantly between systems are denoted the same capital letter.

[6]LSD values at 5% level for the total yields (sum of two cuts) and that for clover content and nutrient concentrations of first cut.

[7]LSD values at 5% level for clover content and nutrient concentrations of the second cut.

The annual DM yield of the systems averaged over crops appeared to follow the same pattern between years within each production group (Figure 2).

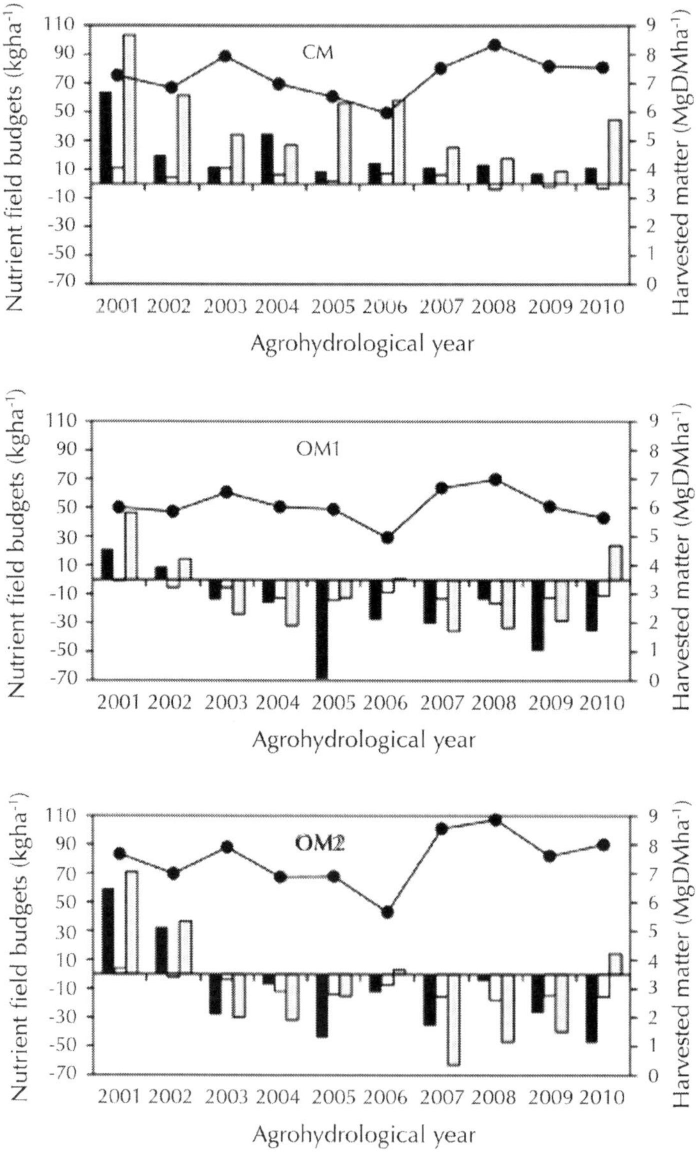

Figure 2: Annual (agrohydrological year) harvested dry matter yields averaged over crops (lines) (right y-axis) and soil system budgets for N (black bars), P

(white bars), and K (grey bars) (left y-axis) for CA1 (conventional arable: upper left subplot), CA2 (conventional arable, environmentally sound: middle left subplot), OA (organic arable: lower left subplot), CM (conventional mixed dairy: upper right subplot), OM1 (organic mixed dairy: middle right subplot), and OM2 (organic mixed dairy with 75% clover ley: lower right subplot).

Nutrient Concentrations of the Harvested Crops

The only differences in nutrient concentrations among the cash crops were for barley N and K (Table 3). The barley N concentration was highest in CA1, followed by CA2 and OM2 whereas OA had the lowest N concentration. The differences were smaller for K, with highest concentration in barley from OM2, and lowest in barley from CA2 and CM.

The N concentration in herbage (grass clover ley) differed significantly between systems at the 2nd cut in both ley year 1 and 2 (ley year 3 was not comparable between systems), with lower concentration in CM than in the two organic systems, which had similar concentrations (Table 4). For the concentrations of K, the tendency was opposite, at least at the first cut. Organically cropped grass clover ley had significantly lower K concentration than that of CM at the first cut of ley year 2

Soil System Nutrient Budgets

Nitrogen

The N input was in the range of 60–112% of the N output (Table 5). The arable system CA1 and all the organic systems had negative soil system N budgets, indicating depletion of the soil organic N content. Over the 10 years, the reductions amounted to 280, 319, 225, and 114 kg N ha^{-1} for CA1, OA, OM1, and OM2, respectively. By contrast, the budget of CA2 appeared to be balanced, whereas CM had an N surplus amounting to 198 kg N ha^{-1}.

Table 5: Measured and estimated nitrogen flows (kg N ha^{-1} $year^{-1}$) and selected quotients (kg N kg N^{-1}) in the cropping systems at Apelsvoll, mean 2001–2010[1]

N flow	Cropping system[2]					
	CA1	CA2	OA	CM	OM1	OM2
Fertilizer applied	124.9	118.6	0.0	87.2	0.0	0.0
Cattle slurry applied	0.0	0.0	0.0	68.8	57.5	82.4
Dry atmospheric depositions	2.7	2.7	2.7	2.7	2.7	2.7
Wet atmospheric depositions	7.2	7.2	7.2	7.2	7.2	7.2
N fixation by legumes	0.0	0.0	34.0	14.3	57.4	59.1
N in seeds	12.9	13.1	4.4	2.5	2.6	1.1
Sum field N input	147.7	141.6	48.3	182.6	127.5	152.6
Harvest[3]	117.4	99.7	43.5	118.4	111.7	130.5
NH_3 and from applied fertilizer[4]	12.5	11.9	0.0	8.7	0.0	0.0
NH_3 and from applied cattle slurry[4,5]	0.0	0.0	6.7	13.8	11.5	16.5
Direct N_2O losses[4]	1.6	1.7	0.7	2.1	1.0	1.1
Drainage and runoff N	44.2	26.7	29.3	19.9	25.6	15.9
Sum field N output	175.7	140.0	80.2	162.8	149.9	164.0
Field N budget	−28.0	1.6	−31.9	19.8	−22.5	−11.4
Harvested N[6]/applied N	0.83	0.84	—	0.73	1.86	1.54
Drainage + runoff N / applied N	0.35	0.23	—	0.13	0.45	0.19
Drainage + runoff N / harvested N[6]	0.43	0.27	0.67	0.17	0.24	0.13

[1]Timestep used is the agrohydrological year (May–April), thus, covering the period May 2001–April 2011.

[2]Each system covers 0.18 ha and consists of four rotation plots á0.045 ha.

[3]Including the N removed with straw. Harvested grass-clover N was reduced by 10% to account for likely harvest-related losses under practical conditions (see Section 2).

[4]Calculated according to [24].

[5]Volatilization of NH_3-N from mulched clover grass is not accounted for in [24], although it may be substantial [25]. It was, therefore, calculated in accordance with Korsaeth [6]. [6]Not including N removed with straw

The annual soil system N budgets (and those for P and K) were rather consistent over years for the arable systems, whereas the annual budgets of the mixed dairy systems appeared to be more positive in 2001 and 2002 than during the rest of the decade (Figure 2).

The amount of N in harvested cereals and potatoes of CA1 and CA2 corresponded to 83 and 84% of that applied, respectively (Table 5). The proportion was somewhat lower in the mixed dairy system CM (73%), whereas in the organic systems OM1 and OM2 the N removal exceeded that applied by 54 and 86%, respectively.

When comparing the sum of N lost via drainage and runoff with that applied (in fertilizer and/or cattle slurry), CA1 had the largest quotient within the arable cropping system group, and, correspondingly, OM1 had the largest quotient within the group of mixed dairy systems (Table 5).

The arable systems had the largest losses of N (via drainage and runoff) per unit of harvested N (loss-to-harvest ratio), with OA having the overall largest loss-to-harvest ratio (Table 5). The water based N losses from this system corresponded to 67% of the N in harvested products. By contrast, CA2 lost only amounts corresponding to 27% of the harvested N. The differences were much smaller within the group of mixed dairy systems, with N losses ranging from 13–24% of the harvested N

Phosphorus

The P input was in the range of 8–156% of the P output (Table 6). The P budgets differed markedly between organic and conventionally managed systems. All three organic systems showed depletions of the soil P content amounting to 82, 100, and 98 kg P ha−1 for OA, OM1, and OM2, respectively. The conventional systems all had a calculated P surplus, particularly CA1 and CA2, which appeared to accumulate P in the same order of magnitude as the P reductions in the organic systems.

Table 6: Measured and estimated phosphorus flows (kg P ha^{-1} $year^{-1}$) and selected quotients (kg P kg P^{-1}) in the cropping systems at Apelsvoll, mean 2001–2010[1]

P flow	Cropping system[2]					
	CA1	CA2	OA	CM	OM1	OM2
Fertilizer	29.8	29.5	0.0	15.0	0.0	0.0
Cattle slurry applied	0.0	0.0	0.0	9.6	7.9	10.9
P in seeds	2.6	2.7	0.7	0.5	0.5	0.2
Sum field P input	32.5	32.1	0.7	25.1	8.4	11.1
Harvested P[3]	24.1	20.5	8.8	21.0	18.2	20.7
Drainage and runoff P	0.2	0.1	0.2	0.3	0.2	0.2
Sum field P output	24.3	20.7	8.9	21.2	18.4	20.9
Field P budget	8.2	11.5	−8.2	3.9	−10.0	−9.8
Harvested P[4]/ applied P	0.72	0.70		0.82	2.17	1.82

[1]Timestep used is the agrohydrological year (May–April), thus, covering the period May 2001–April 2011.

[2]Each system covers 0.18 ha, and consists of four rotation plots á 0.045 ha.

[3]Including the P removed with straw. Harvested grass-clover P was reduced by 10% to account for likely harvest-related losses under practical conditions (see Section 2).

[4]Not including P removed with straw.

The P removal at harvest amounted to 70 and 72% of that applied in CA1 and CA2, respectively (Table 6). This percentage was somewhat larger for CM (82%), and very high for the organic systems OM1 (217%) and OM2 (182%), indicating soil P mining in these systems.

As the system differences in terms of P losses in drainage and runoff were not statistically significant, quotients between these losses and applied or harvested P were not calculated

Potassium

The K input was in the range of 23–200% of the K output (Table 7). The pattern of the K budgets was similar to that for P; all the organic systems had calculated K deficits, whereas all the conventional systems appeared to accumulate K.

Table 7: Measured and estimated potassium flows ($kg\,K\,ha^{-1}\,year^{-1}$) and selected quotients ($kg\,K\,kg\,K^{-1}$) in the cropping systems at Apelsvoll, mean 2001–2010[1]

K flow	Cropping system[2]					
	CA1	CA2	OA	CM	OM1	OM2
Fertilizer	90.7	88.4	0.0	56.9	0.0	0.0
Cattle slurry applied	0.0	0.0	0.0	89.2	74.7	101.8
Wet atmospheric depositions	2.1	2.1	2.1	2.1	2.1	2.1
K in seeds	11.4	11.4	0.9	0.6	0.6	0.3
Sum field K input	104.2	101.9	3.0	148.8	77.4	104.2
Harvested K[3]	70.4	46.8	9.3	101.4	82.2	112.0
Drainage and runoff K	6.6	4.1	3.3	3.6	3.3	2.3
Sum field K output	77.0	50.9	12.6	105.1	85.4	114.3
Field K budget	27.2	51.0	−9.7	43.7	−8.0	−10.1
Harvested K[4]/applied K	0.53	0.53		0.66	1.04	1.05
Drainage + runoff K/ applied K	0.07	0.05		0.03	0.04	0.02
Drainage + runoff K/ harvested K[4]	0.14	0.09	0.36	0.04	0.04	0.02

[1]Timestep used is the agrohydrological year (May–April), thus covering the period May 2001–April 2011.

[2]Each system covers 0.18 ha and consists of four rotation plots á0.045 ha.

[3]Including the K removed with straw. Harvested grass-clover K was reduced by 10% to account for likely harvest-related losses under practical conditions (see Section 2).

[4]Not including K removed with straw

The amount of harvested K corresponded to 53% of that applied in CA1 and CA2 (Table 7). This proportion was somewhat higher in the mixed dairy system CM (66%). In the organic systems OM1 and OM2 the K removal at harvest was only slightly larger than the amount of K applied.

The sum of K lost via drainage and runoff was 7% and 5% of that applied in CA1 and CA2, respectively (Table 7). The corresponding percentages for the mixed dairy systems were less, ranging from 2–4%.

The loss-to-harvest ratios for K followed a similar pattern as for N, but at a lower level (Table 7). The arable systems had the largest ratios, with the highest value calculated for OA. The differences were much smaller within the group of mixed dairy systems, with N losses ranging from 2–4% of the harvested K

Changes in Topsoil Nutrient Content

Nitrogen

The ignition loss did not differ significantly between 1999 and 2009, although the measurements appeared to be at a lower level in 2009 (Table 8, Figure 3(a)). The tendency of reduction was particularly strong for CA1 (p= 0.0.054) and OM2 (0.088). Although the measured changes in ignition loss over time did not support the calculated N budgets substantially, the differences between systems in terms of ignition loss measured in 2009 reflected the N budgets very well (Figure 4(a)). A ranking of the systems in terms of measured ignition loss levels in 2009 was almost identical with a ranking based on the calculated N budgets

Table 8: Measured changes in selected chemical properties in topsoil (0–25 cm) for the last decade

Cropping system/year	CA1	CA2	OA	CM	OM1	OM2
	Ignition loss ($g\,100\,g^{-1}$)					
1999	6.83	6.88	7.05	7.30	6.98	7.13
2009	6.13	6.70	6.30	6.78	6.50	6.63
Change (2009–1999)	−0.70	−0.18	−0.75	−0.53	−0.48	−0.50
value[1]	0.054[n.s.]	0.608[n.s.]	0.204[n.s.]	0.152[n.s.]	0.265[n.s.]	0.088[n.s.]
	P-AL ($mg\,100\,g^{-1}$)					
1996	5.38	4.00	4.25	4.00	3.75	3.88
2009	7.10	8.23	3.13	6.00	3.00	3.65
Change (2009–1996)	1.73	4.23	−1.13	2.00	−0.75	−0.23
value	0.029	0.015	0.036	0.001	0.102[n.s.]	0.409[n.s.]
	K-AL ($mg\,100\,g^{-1}$)					
1996	6.75	6.75	6.13	7.13	6.13	6.00
2009	6.73	12.0	6.35	9.63	6.30	5.55
Change (2009–1996)	−0.03	5.28	0.23	2.50	0.18	−0.45
value	0.942[n.s.]	0.043	0.362[n.s.]	0.022	0.188[n.s.]	0.517[n.s.]
	K-HNO_3 ($mg\,100\,g^{-1}$)					
1996	31.1	28.3	30.9	28.6	29.1	26.0
2009	47.8	58.5	39.3	55.5	39.0	35.3
Change (2009–1996)	16.6	30.3	8.38	26.9	9.88	9.25
value	0.012	0.007	0.010	0.003	0.001	0.037

[1]Level of significance for the pairwise -test of differences. Non significance at the 5% level is denoted[n.s]

Figure 3: Topsoil (0–25 cm) content of ignition loss (subplot A), P-AL (subplot B), K-AL (subplot C), and K-HNO_3 (subplot D) for the six cropping systems CA1 (conventional arable: ○), CA2 (conventional arable, environmentally sound: □), OA (organic arable: Δ), CM (conventional mixed dairy: •), OM1 (organic mixed dairy: ▲), and OM2 (organic mixed dairy with 75% clover ley: ■)

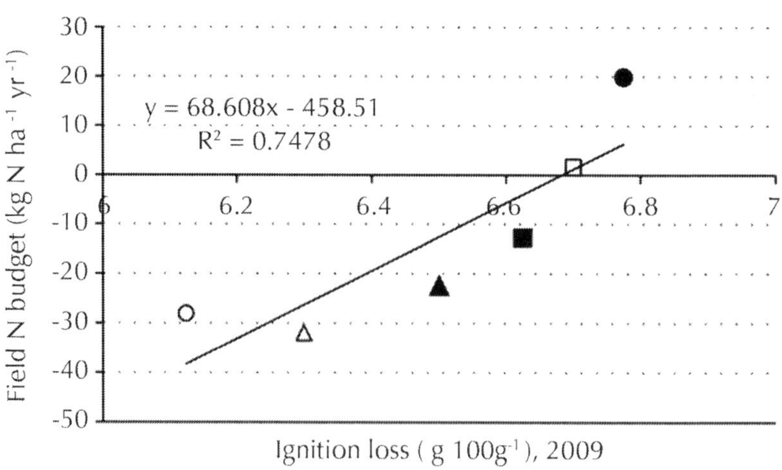

Figure 4: Calculated soil system budgets of N, P, and K for the years 2001–2010 plotted, respectively, against ignition loss (subplot A), P-AL (subplot B),

K-AL (subplot C), and K-HNO_3 (subplot D) for the six cropping systems CA1 (conventional arable: ○), CA2 (conventional arable, environmentally sound: □), OA (organic arable: Δ), CM (conventional mixed dairy: ●), OM1 (organic mixed dairy: ▲), and OM2 (organic mixed dairy with 75% clover ley: ■).

Phosphorus

P-AL changed significantly in four of the six systems during the period 1996–2009 (Table 8, Figure 3(b)). The measurements in 2009 showed the same pattern of differences between the systems as did those in 2003, but with a greater magnitude. The measured differences in P-AL reflected the calculated P budgets well, although the measured declines in OM1 and OM2 were not significant (Table 8). There was a strong linear relationship between P-AL measured in 2009 and the calculated P budgets of the systems (Figure 4(b)).

Potassium

K-AL followed a pattern similar to that of P-AL over the period 1996–2009, with increasing differences between systems over time (Figure 3(c)). The only significant differences between the measurements in 1996 and 2009 were found in CA2 and CM, which both showed increased levels of K-AL in 2009 (Table 8). The calculated K-deficits in the organic systems, and the K-surplus of CA1, could, thus, not be supported by the measured differences in K-AL over time. The differences between systems in terms of K-AL measured in 2009 corresponded, however, very well with the calculated K budgets (Figure 4(c)). There was a significant increase in the topsoil content of K-HNO_3 from 1996 to 2009 in all systems (Table 8, Figure 3(d)). The rate of change decreased in the order CA2>CM> CA1>OM1>OM2>OA, a ranking which corresponded well with a ranking of the calculated K budgets. The relation between the soil measurements in 2009 and the calculated K budgets was even stronger for K-HNO_3 than it was for K-AL (Figure 4(d)).

DISCUSSION

Yields

Both cereal yield levels and relative yield differences between systems followed the same pattern for the whole period 2001–2010 as for the years 2001–2004, which were discussed by Korsaeth [6]. Briefly, the yield differences were larger amongst the arable systems than amongst the mixed dairy systems. The low yields in OA may be explained by P and K deficits, as indicated by the negative P and K budgets and the significant reduction in plant available topsoil P, and due to the lack of chemical plant protection. More foliar diseases were generally observed on cereals in the organic systems than in the other systems (data not shown). The lack of significant differences in measured potato parameters between the two cropping systems (CA1 and CA2) is partly in accordance with the findings of Riley and Ekeberg [27], who compared spring and autumn ploughing at different depths (10, 20, and 30 cm) with tine harrowing only in spring, on the same soil type at a nearby location. They found the same potato fresh weight yields in all treatments, but the tuber dry matter concentration was significantly lower in potatoes grown without ploughing.

Among the mixed dairy systems, the organically grown cereal yields were also lower than those grown conventionally. However, due to much smaller differences in nutrient regime, the differences between mixed dairy systems were less than those between arable systems. The most likely reasons for inferior yields in the organic mixed dairy systems relative to the conventional mixed dairy system were, as for the arable systems, suboptimal nutrition and a lack of plant protection. The yield pattern of grass clover ley was also unchanged in 2005–10 compared with the first four years of the decade [6]. The high yields of the organic leys may be explained partly by their N fixation, which was estimated to be much greater in the organic leys than in the conventional leys. This was a result of the significantly higher proportions of clover in the organic leys and less suppression of N fixation by the use of inorganic N fertilizer. The reduced grass clover yields in the 3rd ley year of OM2 may be due to the reduced proportion of clover compared with the first two ley years.

The long period of active nutrient uptake by grass and clover may also partly explain the relatively high organic ley yields. A longer uptake period increases the utilization of less readily available nutrients (e.g., nutrients in organic form), since the mineralization of such nutrients occurs throughout the cropping season. Smaller yield differences between organic and conventional cropping for grass clover ley than for cereals have been reported previously for this experiment [28].In a review of a number of Swedish field studies, Bergström et al. [29] reported that crop yields in organic rotations were reduced by 20 to 80%, compared with the same crops in conventional rotations. These authors explained this in terms of higher N deficiency, more weed competition, and greater infestation of crop diseases in the organic systems.

Soil System Nutrient Budgets and Nutrient Concentrations of Topsoil and Crops

Nitrogen

The large calculated deficits found for the arable systems CA1 and OA in the first part of the decade [6] were sustained. The suggested net soil N depletion corresponded to a relative decay rate of the topsoil (0–25 cm) N content of 0.4% yr^{-1}. This corresponds well with Riley and Bakkegard [30], who compared soil samples taken in 1991 and in 2001 from 291 arable fields located throughout southeast Norway. They found that the percentage relative decline rate of SOM was approximately one tenth of the initial percentage of organic matter in soil over the decade. In the present study, there was a strong tendency ($P = 0.054$) towards reduced ignition loss for CA1 in 2009 compared with 1999; but this was not the case for OA.

Comparing conventional and organic cropping systems in a pipe-drained plot experiment in Sweden, Torstensson et al. [31] also reported an N deficit (−18 kg N ha^{-1} yr^{-1}, not including denitrification, N in seeds and atmospheric depositions) for a conventional arable rotation (CON, barley-oat-spring wheat-barley-oat-potato) comparable with CA1. They tested additionally an arable organic system with green manure as the only N source (OGM, oat-green manure-spring wheat-oat-green manure-potato), comparable with OA and found a positive soil system N budget (13 kg N ha^{-1} yr^{-1}) in contrast to the present findings. In the experiment of Torstensson et al. [31], the proportion of green manure was, however, larger than in our case (33% versus 25%), which resulted in 37 kg N ha^{-1} yr^{-1} more N fixation and 20 kg N ha^{-1} yr^{-1} less harvested N, compared with OA.

The only arable system which appeared to have a balanced N budget was CA2. This may mainly be explained by its comparative low leaching and runoff N losses, which represent the main difference between CA1 and CA2 with regard to N flows. The findings indicate that reduced tillage counteracted soil N mining, which is a commonly reported result (e.g., [32]). Another factor which may have contributed to prevent soil N mining in CA2 is that straw was not removed. Straw incorporation has a well-known positive effect on the soil organic N content (e.g., [33]) The conventional mixed dairy system CM had a calculated N surplus over the decade, which indicates that the system probably increased its soil organic matter content. Conservation of or an increase in soil N has also been reported for other rotations containing pasture or ley receiving organic N on relatively N-rich (>2.0 g kg^{-1}) soils [32, 33]. The opposite was found in the organic system OM1, with the same crop rotation and tillage as CM, but with a calculated N deficit of 23 kg N ha^{-1} yr^{-1}, indicating that the relatively high production has been maintained at the cost of the soil organic N pool. Similarly, Steinshamn et al. [34] reported an annual N deficit of 16 kg N ha^{-1}, not including N leaching and denitrification, at the field level in an organic crop rotation with 50% grassland (barley, forage rape + Italian ryegrass, oats + peas, 3-year grassland). The slightly negative N budget of OM2 (−16 kg N ha^{-1}) shows that a high proportion of grass clover ley in the rotation does not guarantee a balanced N budget. By comparison, Syväsalo et al. [35] reported an even larger N deficit (−31 kg N ha^{-1} yr^{-1}, not including ammonia emissions, deposition, or N in seeds) in an organic grass clover ley receiving 130 kg N ha^{-1} in cattle slurry. An extra year of grass clover ley instead of wheat in the OM2 rotation resulted in the largest calculated amounts of available cattle slurry for this system (82 kg N ha^{-1}), but the N fixation was apparently much less effective in the additional ley year than in the previous two ley years. The estimated N fixation of the third ley year was only 43 kg N ha^{-1} yr^{-1}, compared with 82 and 88 kg in the 1st and 2nd ley year, respectively. The annual nutrient budgets of the mixed dairy systems showed more positive figures for 2001 and 2002 compared with the following eight years. This was a result of too high yield expectations when calculating the initial amount of slurry available for these systems in 2001. The number of cows which each farming system could sustain, and, thus, the amount of slurry available for the crops, was calculated from the average total available feed in the system during the previous three years (sliding mean). The initialization problem was, thus, gradually levelled out. The calculated changes in soil N, that is, misbalanced soil system N budgets, were in general poorly supported by the measured changes in topsoil ignition loss since 1999, which were all nonsignificant. Considering the large differences in the measured N flows between the systems, it is very unlikely that the SOM level of 1999 would have been sustained in all systems over the following decade. One explanation of the mismatch could be an over- or underestimation of the estimated gaseous N losses, which were the most uncertain N flows in the calculated budgets. If these losses were largely overestimated, the calculated deficits of CA1, OA, OM1, and OM2 would have been reduced, but the calculated surplus of OM would have increased, and vice versa. Another explanation could be that some organic matter has been transported from topsoil to subsoil. Such a translocation of organic matter may have taken place, but it seems unlikely that this process has diverged significantly between systems with more or less the same crop rotation.

The relative differences between the systems in 2009 matched the calculated soil system N budgets much better than the differences found between sampling times, indicating that the 1999 data may include some random variation. The relation between ignition losses in 2009 and the N budgets indicated that ignition loss would equilibrate at 67 $g\,kg^{-1}$ with a balanced N budget. This corresponds to an SOM content of 47 $g\,kg^{-1}$, calculated with a pedotransfer function developed solely for this site (SOM = 0.81 × ignition loss (%)−0.038 × clay (%)−0.70), [21]). The significant differences between systems in terms of N concentrations of the harvested crops were few and reflected poorly the differences in fertilization regime and soil system N budgets. Greenwood et al. [36] developed a model linking N concentration in plant DM to growth rate and to plant mass per unit area. They found that subcritical values of N concentration during growth affected the growth rate. In order to define whether any of the crops in the present study were N limited, measurements of N concentration during crop growth and not at harvest would, thus, have been required.

Phosphorus

Conventional arable cropping appeared to give relatively large surpluses of P. In the past, more emphasis, in Norway, was placed on adjusting N fertilizer rates to crop requirements than to adjusting P rates. In 2007–2008 there was a change in the fertilizer recommendations for P in Norway, with a reduction of approx. 25% for cereals and grasses and 30% for potatoes. This prompted the fertilizer market leader in Norway (market share > 90%) to increase the N:P-ratio of their most used compound fertilizers, from 5.3 to 7.3 for cereals and from 2.2 to 3.0 for potatoes, thus reducing the amounts of P for a given amount of N.The organic arable system did not receive any P, and the production was thus entirely dependent on P supply from the soil. The negative soil system budget of 8 kg P ha−1 yr^{-1} was the same as that found in a stockless organic farm (red clover-winter wheat-spring beans-spring cereal) in the UK [15]. The results show that in stockless organic farming systems, some form of external P addition becomes unavoidable sooner or later (depending on the size of the initial P pool and the ability of the soil to deliver plant available P). Berry et al. [15] showed that a system comparable with OA was almost in balance in terms of P when it received rock phosphate. From a resource economics point of view, it is questionable; however, whether the use of untreated rock phosphate is a good strategy, considering its low plant availability [37]. An alternative could be to use organic waste, such as biogas residue from household waste, which has been shown to be a valuable and inexpensive source of plant nutrients [38]. The conventional mixed dairy system had a field surplus of almost 4 kg P ha^{-1} yr^{-1}, suggesting an unnecessary use of a limited resource. This appears to be no exception. In a comparable farming system in northern Sweden, Bengtsson et al. [39] reported a P surplus of 5 kg P ha^{-1} yr^{-1}. The problem seems to be even worse on conventional dairy farms, that is, those with no or a very low proportion of arable crops. The P surplus on such farms is assumed to vary from 10 to 72 kg ha^{-1} in Europe (Pfimlin et al. 2006, cited by [40]). The organic mixed dairy systems produced at the cost of their indigenous soil P pool, with a total deficit of about 10 kg P ha^{-1} yr^{-1}. This was not surprising, considering that there was no P input to these systems, except for that in seeds. Even when some feed is purchased, P deficits are commonly reported. Berry et al. [15] reported soil surface budgets (i.e., maximum root depth as lower boundary) of −3 kg P ha^{-1} yr^{-1}

in a mixed dairy system in the UK (ley cropped in 3 out of 5 years, farm number 3). Steinshamn et al. [34] found a deficit between inputs and produce (losses not considered) in an organic dairy farming system in Norway of 6.3 kg P ha^{-1} yr^{-1}. All the significant changes in topsoil P-AL from 1996 to 2009 were in the same direction as the corresponding calculated soil system P budgets. The budget calculations were also supported by the very strong relation between the topsoil level of AL-extractable P in 2009 and the calculated soil system budget. The results clearly show the effect of both overfertilization and suboptimal P fertilization on the plant availability of P in soil. In the present study, P losses to subsoil, that is, below root depth, were considered to be negligible. On a comparable soil in the long-term fertilizer trials at Møystad in S.E. Norway, Riley [41] found no effect of P treatment (no P addition, P in mineral fertilizer, or P in animal manure) below 40 cm depth. In contrast, Verloop et al. [40], studying intensive dairy farming systems, found that some topsoil P was transported to subsoil and that the P accumulation in the deeper layers practically equalled the P depletion in the upper topsoil. Their experiment was, however, run on a light sandy soil, characterized by a 0.3 m anthropogenic topsoil overlaying a layer of yellow sand hardly penetrable by roots. With an assumed balanced P budget, P-AL appeared to equilibrate at 54 mg P kg^{-1}, a level which is normally considered adequate for optimum growth. In the long-term trials at Møystad, Ekeberg and Riley [8] also found a strong relationship between topsoil P-AL and P balance (P applied via fertilizer and/or farmyard manure minus P removed by harvest) for the period 1922–1983. They reported that P-AL equilibrated at 25–30 mg P kg^{-1} when the application of P equalled the removal of P by crops. This equilibrium point rose to about 40 mg P kg^{-1} in the period 1983–2003 (H. Riley, personal communication). Experiences from an intensive dairy farm in The Netherlands, where P-equilibrium fertilization (i.e., balancing P inputs via fertilizer and manure with P in crop products) is performed, has shown that the soil available P-status differs between crop rotations. The P content of the crops did not differ between the systems, and the herbage concentrations of P were in the range of 0.2–0.3%, a level which is regarded as adequate [42]. In comparison, Mathews et al. [43] considered 0.2–0.34% P to be the critical concentration for cool-season grasses, that is, a concentration level below which a 10% yield drop is expected. It, thus, appears that the grass clover growth was not P limited in the organic mixed dairy systems, in spite of the continuous soil P depletion of these systems.

Potassium

The potassium soil system budgets showed that the conventional arable systems had unnecessarily high levels of K fertilization. As for P, the change in the compound fertilizers (from 2009) also altered the amount of K relative to N, with a N:K-ratio increasing from 2.1 to 2.2 for cereal fertilizer and from 0.65 to 0.67 for potato fertilizer. Although this change has some importance for practical farming, it did not influence the results presented here.

In contrast to the present findings, Torstensson et al. [31] reported a small K deficit (-3 kg K ha^{-1} yr^{-1}) in a 6 year conventional rotation with five years of spring cereals and one year with potatoes, comparable with CA1. Additionally, they found a large deficit (-28 kg K ha^{-1} yr^{-1}) in a similar crop rotation but with ryegrass grown as a catch crop after each main crop, comparable with CA2. The contrasting findings of Torstensson et al. [31] may largely be explained by the K leaching, which were 5–7 times larger in their experiment than in the present study. The literature appears inconclusive, however, when it comes to K-budgets for conventional arable cropping systems. This was well illustrated by Heming [44], who studied a large number of fields in southern England, reporting K budgets (K applied in fertilizer minus K in crop) ranging from -40 to $+70$ kg K ha^{-1} yr^{-1}.

The organic arable system OA had a calculated K-deficit. Stockless organic systems without any form of K application are bound to result in negative K budgets, as has been commonly reported (e.g., [15, 31, 45]). Interestingly, the calculated deficit in OA was of the same magnitude as the calculated deficits of the two organic mixed dairy systems (OM1 and OM2). The relatively large inputs of K in applied slurry in the mixed dairy systems, thus, appeared to be more than outweighed by large K export via harvested material. Reviewing a range of cropping systems in northern Europe, Öborn et al. [46] summarized that negative farm gate and soil surface K budgets are especially common in organic farming. This was supported by Øgaard and Hansen [47], who found negative K budgets in grassland fields on 23 out of 26 organic farms in Norway. In a study of three long-term field experiments with mixed cropping systems (six year rotations with 2/6 or 3/6 ley) on sandy loam soils over the 18 year period 1997–2004, Andrist-Rangel et al. [48] reported negative K budgets (input minus crop offtake) for organic systems in the range of -22 to -75 kg K ha^{-1} yr^{-1}. They also found, however, that a conventional mixed cropping system had negative field budgets during the same period, ranging from −21 to −60 K $ha^{-1} yr^{-1}$, in contrast to the conventional mixed cropping system of the present study (CM), which had a calculated surplus of 44 kg K $ha^{-1} yr^{-1}$. On the other hand, Bengtsson et al. [39] reported a K surplus of 39 kg K $ha^{-1} yr^{-1}$ in a comparable conventional system in northern Sweden. The only significant changes in measured topsoil K-AL were the increase in CA2 and CM, in correspondence with their large calculated field K surpluses. The relative differences between systems in terms of calculated K budgets were well reflected in the relative differences between changes of K-HNO_3 and there was a clear relation between K budgets and measured levels of both K-AL and K-HNO_3 in 2009. The relatively large, negative soil system K budgets for the organic systems were, however, not reflected by the soil measurements. Could it be that the soil system K budgets were largely underestimated (i.e., underestimated inputs and/or overestimated outputs)?

Errors in K mass budgets are not an uncommon phenomenon (e.g., [46]). It seems unlikely, however, that such errors alone could explain the lack of fit between the calculated K budgets and the soil K measurements. For example, the unfertilized system OA would need a budget correction of about 10 kg K $ha^{-1} yr^{-1}$, in order to achieve a balanced soil system K budget, which would reflect the K-AL data (i.e., no significant difference between 1996 and 1999). Such a change would be of the same magnitude as the entire (current) calculated K off-take by harvest, and it would correspond to more than 330% of the total K input or 300% of the K measured in drainage and runoff. Moreover, the calculated depletions of K in the organic systems are well supported by relevant comparisons in the literature, as discussed above.

Another, and more likely, explanation of the poor relation between soil system K budgets and changes in soil K measurements over time, could be that the weathering of primary minerals has released K in amounts compensating for the calculated K depletion. Annual weathering rates for different Norwegian and Swedish soils have been estimated to range from 3 to 82 kg K ha^{-1} [49]. Øgaard and Hansen [47], looking at K uptake and requirement in organic grassland farming in Norway, found that K uptake from reserve K, that is, K located in the interlayers of the sheet silicates in the soil, was positively related to the acid soluble K content (i.e., K-HNO_3 minus K-AL) of the soil. In the present study, acid soluble K was in the range of 20–25 mg K 100 g^{-1} (1996 values), which corresponds to a potential uptake of reserve K between 30 and 110 kg K ha^{-1} yr^{-1}, based on the findings of Øgaard and Hansen [47].

The considerations above relate to the topsoil. Additionally, there may have been substantial K uptake from the subsoil, particularly in systems with negative soil system budgets. Experiments using a K/Rb isotope dilution method on loess-parabrown soils in N. Germany have showed that the subsoil (>30 cm depth) supplied between 9 and 70% of the total K uptake in spring wheat [50].

The data give no reason to assume that the availability of K was limiting for yield formation. In the future, separate analyses of grasses and legumes will nevertheless be performed, to enable a better assessment of the critical nutrient levels of the forage crops.

Implications for Future Cropping Systems

Agronomic practices affect the balance between utilization and removal of plant available nutrients in soil. Where the reserves of potentially available nutrients (e.g., organic N, organic/fixed P, and fixed K) are large, low input farming systems may maintain productivity at the cost of a gradual decline of the soil nutrient pools. It should be emphasised, however, that soil P mining is to be considered as an equally serious depletion of a limited resource as the mining of rock P for producing mineral fertilizer. In future agricultural systems, one main challenge is to close the P cycle. Today, a large share of harvested P ends up in persistent chemical compounds due to the widespread use of coagulants (e.g., aluminium and iron) to remove P from sewage [51]. One step in the right direction is to use biogas residue from household waste, as mentioned above, as a nutrient source. Such a practice has from 2011 been integrated into the management of the organic OA system of the current study, in order to improve its nutrient balance. Nitrogen may be fixed biologically in sufficient amounts in clover ley dominated systems. Systems which have an N input based on green manure appear, however, to be very area-demanding, and the practice of leaving large amounts of N-rich material unharvested leads to severe risks for N losses to the environment [6]. The present study shows that a good alternative is to use moderate amounts of fertilizer, combined with steps to reduce the risk of losses, such as the use of spring tillage, catch crops, and split application of fertilizer. Such a system (i.e., CA2) has previously been shown to have the lowest ratio of N loss to food production [6].

Potassium is often given relatively little attention when studying agricultural systems, probably due to the absence of direct, negative environmental impacts associated with K losses. Moreover, many soils show a remarkable capacity for replacing removed plant available K through weathering and, thus, counteracting negative effects (i.e., plant K deficiency) when soil K mining occurs, as shown in this study. In the long term, however, all systems require a balanced supply of K, which implies that most organic farmers today need to consider means to enhance the K input to their farms. This study highlights also the need for balancing K, along with N and P, in conventional systems, as there is a risk for an oversupply of K in these systems. Oversupply means that the indirect environmental impacts, that is, those associated with the manufacturing and transport of K fertilizer, are unnecessarily increased.

To perform a thorough evaluation of agricultural systems in terms of environmental impact, such as their global warming potential, all processes governing the manufacturing and transport of input factors (not only those for fertilizer) should be considered. This points toward the need for life cycle assessment (LCA) studies, which use a holistic approach for the evaluation of a system's environmental impacts (e.g., [52]). An LCA study of the systems included in the present paper is currently in preparation.

CONCLUSIONS

Differences in cereal yields between organically and conventionally managed systems are larger in arable rotations than those in mixed cropping systems including ley and animal husbandry, mainly due to improved nutrition through N fixation in the ley and the availability of animal manure in the latter group.

The yield differences for grass clover leys are smaller between the two management types (i.e., organic versus conventional) than those of cereals, due to lower disease pressure in grasses and better nutrition, resulting from N fixation and better utilization of mineralised nutrients (i.e., long period of active uptake).

Arable cropping may result in soil N mining, even when fertilized with normal to high amounts of N, due to high potential for losses and poor utilization of mineralised N. The use of 25% of the production area for green manure production is insufficient as an N source alone to balance the N losses and the off-take by harvest, on a fertile soil with high yield potential. Arable cropping, which comprises the use of reduced tillage, catch crops, and moderate amounts of fertilizer appears, however, to balance the N flows at field level. The use of P and K fertilizer in arable crop production may be used to balance the respective nutrient flows but it should be used with care, as over supply is a great risk.

Mixed dairy systems, producing both cereals and fodder crops, risk an undersupply of both N, P, and K, if there is no import of these nutrients to the system in the form of purchased fodder and/or other nutrient sources. The fixation of N by the legumes in the forage crops appears to be insufficient as the only N source. The use of mineral fertilizer may very well be used to balance the flows of N, P, and K in mixed dairy systems, but this leads to a risk of oversupply.A relatively fertile soil may be exposed to substantial mining of N, P, and K over many years without causing detectible nutrient deficits in plants cropped on this soil. A long-term over- or undersupply will, sooner or later, result in a significant change in the content of plant available nutrients in the soil, but this change may be masked by the release of nutrients from nonavailable compounds.

Field nutrient budgets appear to be a good approach to the evaluation of whether a system is managed in a balanced way or not with regard to important nutrients such as N, P, and K, long before eventual unbalances become detectible by traditional analyses. However, the approach requires comprehensive datasets, which are usually unavailable under practical conditions.

ACKNOWLEDGMENTS

H. Riley is gratefully acknowledged for critically reading the paper and T. Gaardløs for his skilled technical assistance. The project was funded jointly by the Norwegian Institute for Agricultural and Environmental Research, the Research Council of Norway, and Yara International ASA.

REFERENCES

1. R. Eltun, "The Apelsvoll cropping system experiment. III. Yield and grain quality of cereals,"Norwegian Journal of Agriculture, vol. 10, no. 1, pp. 7–22, 1996.
2. A. Korsaeth and R. Eltun, "Nitrogen mass balances in conventional, integrated and ecological cropping systems and the relationship between balance calculations and nitrogen runoff in an 8-year field experiment in Norway," Agriculture, Ecosystems and Environment, vol. 79, no. 2-3, pp. 199–214, 2000.
3. G. Lien, O. Flaten, A. Korsaeth et al., "Comparison of risk in organic, integrated and conventional cropping systems in Eastern Norway," Journal of Farm Management, vol. 12, no. 7, pp. 385–401, 2006.
4. T. A. Breland and R. Eltun, "Soil microbial biomass and mineralization of carbon and nitrogen in ecological, integrated and conventional forage and arable cropping systems," Biology and Fertility of Soils, vol. 30, no. 3, pp. 193–201, 1999.
5. H. Riley, R. Pommeresche, R. Eltun, S. Hansen, and A. Korsaeth, "Soil structure, organic matter and earthworm activity in a comparison of cropping systems with contrasting tillage, rotations, fertilizer levels and manure use," Agriculture, Ecosystems and Environment, vol. 124, no. 3-4, pp. 275–284, 2008.
6. A. Korsaeth, "Relations between nitrogen leaching and food productivity in organic and conventional cropping systems in a long-term field study," Agriculture, Ecosystems and Environment, vol. 127, no. 3-4, pp. 177–188, 2008
7. B. T. Christensen and A. E. Johnston, "Chapter 18 Soil organic matter and soil quality-Lessons learned from long-term experiments at Askov and Rothamsted," Developments in Soil Science, vol. 25, no. C, pp. 399–430, 1997.
8. E. Ekeberg and H. Riley, "The long-term fertilizer trials at Møystad, S:E: Norway," SP Report 29, 100th Anniversary Workshop Askov Experimental Station, 1995.
9. N. A. Fettell and H. S. Gill, "Long-term effects of tillage, stubble, and nitrogen management on properties of a red-brown earth," Australian Journal of Experimental Agriculture, vol. 35, no. 7, pp. 923–928, 1995.

10. P. R. Poulton, "The importance of long-term trials in understanding sustainable farming systems: the Rothamsted experience," Australian Journal of Experimental Agriculture, vol. 35, no. 7, pp. 825–834, 1995.
11. W. R. Raun, G. V. Johnson, S. B. Phillips, and R. L. Westerman, "Effect of long-term N fertilization on soil organic C and total N in continuous wheat under conventional tillage in Oklahoma," Soil and Tillage Research, vol. 47, no. 3-4, pp. 323–330, 1998.
12. J. Leifeld, R. Reiser, and H. R. Oberholzer, "Consequences of conventional versus organic farming on soil carbon: results from a 27-year field experiment," Agronomy Journal, vol. 101, no. 5, pp. 1204–1218, 2009.
13. C. F. Drury, T. O. Oloya, D. J. McKenney, E. G. Gregorich, C. S. Tan, and C. L. vanLuyk, "Long-term effects of fertilization and rotation on denitrification and soil carbon," Soil Science Society of America Journal, vol. 62, no. 6, pp. 1572–1579, 1998.
14. C. M. Penfold, M. S. Miyan, T. G. Reeves, and I. T. Grierson, "Biological farming for sustainable agricultural production," Australian Journal of Experimental Agriculture, vol. 35, no. 7, pp. 849–856, 1995.
15. P. M. Berry, E. A. Stockdale, R. Sylvester-Bradley et al., "N, P and K budgets for crop rotations on nine organic farms in the UK," Soil Use and Management, vol. 19, no. 2, pp. 112–118, 2003.
16. O. Oenema, H. Kros, and W. De Vries, "Approaches and uncertainties in nutrient budgets: implications for nutrient management and environmental policies," European Journal of Agronomy, vol. 20, no. 1-2, pp. 3–16, 2003.
17. O. Oenema and M. Heinen, "Uncertainties in nutrient budget due to biases and errors," in Nutrient Disequilibria in Agroecosystems: Concepts and Case Studies, E. M. A. Smaling, O. Oenema, and L. O. Fresco, Eds., pp. 75–97, CAB International, Wallingford, UK, 1999.
18. W. De Vries, J. Kros, O. Oenema, and J. De Klein, "Uncertainties in the fate of nitrogen II: a quantitative assessment of the uncertainties in major nitrogen fluxes in the Netherlands," Nutrient Cycling in Agroecosystems, vol. 66, no. 1, pp. 71–102, 2003.

19. A. Korsaeth, "Runoff and leaching losses of N, P, and K in the period 2001–2011 in a long-term experiment with conventional and organic crop rotations," In press.
20. WRB, World Reference Base for Soil Resources, FAO, Rome, Italy, 1998.
21. H. Riley and R. Eltun, "The Apelsvoll cropping system experiment. II. Soil characteristics," Norwegian Journal of Agricultural Sciences, vol. 8, pp. 317–333, 1994.
22. H. Egnér, H. Riehm, and W. R. Domingo, "Untersuchungen über die chemische Bodenanalyse als Grundlage für die Beurteilung de Nährstoffzustandes der Böden. II," Kungliga Lantbrukshögskolans Annaler, vol. 26, pp. 199–215, 1960.
23. W. Aas, S. Solberg, S. Manø, and K. E. Yttrei, "Overvåking av langtransportert forurenset luft og nedbør," Report NILU OR 22/2007, Kjeller, Norway, 2007.
24. IPCC National Greenhouse Gas Inventories Programme, "Agriculture, forestry and other land use," in IPCC Guidelines for National Greenhouse Gas Inventories, H. S. Eggleston, L. Buendia, K. Miwa, T. Ngara, and K. Tanabe, Eds., vol. 4, chapter 11, pp. 1–11, IGES, Japan, 2006.
25. L. Larsson, M. Ferm, Å. Kasimir-Klemedtsson, and L. Klemedtsson, "Ammonia and nitrous oxide emissions from grass and alfalfa mulches," Nutrient Cycling in Agroecosystems, vol. 51, no. 1, pp. 41–46, 1998.
26. K. A. Gomez and A. A. Gomez, Statistical Procedures for Agricultural Research, Wiley & Sons, New York, NY, USA, 2nd edition, 1984.
27. H. Riley and E. Ekeberg, "Effects of depth and time of ploughing on yields of spring cereals and potatoes and on soil properties of a morainic loam soil," Acta Agriculturae Scandinavica - Section B Soil and Plant Science, vol. 48, no. 4, pp. 193–200, 1998.
28. R. Eltun, A. Korsaeth, and O. Nordheim, "A comparison of environmental, soil fertility, yield, and economical effects in six cropping systems based on an 8-year experiment in Norway," Agriculture, Ecosystems and Environment, vol. 90, no. 2, pp. 155–168, 2002.

29. L. Bergström, H. Kirchmann, H. Aronsson, G. Torstensson, and L. Mattsson, "Use Efficiency and Leaching of Nutrients in Organic and Conventional Cropping Systems in Sweden," in Organic Crop Production—Ambitions and Limitations, H. Kirchmann and L. Bergström, Eds., pp. 117–141, Springer, New York, NY, USA, 2008.
30. H. Riley and M. Bakkegard, "Declines of soil organic matter content under arable cropping in southeast Norway," Acta Agriculturae Scandinavica Section B, vol. 56, no. 3, pp. 217–223, 2006.
31. G. Torstensson, H. Aronsson, and L. Bergstrom, "Nutrient use efficiencies and leaching of organic and conventional cropping systems in Sweden," Agronomy Journal, vol. 98, no. 3, pp. 603–615, 2006.
32. D. P. Heenan, W. J. McGhie, F. M. Thomson, and K. Y. Chan, "Decline in soil organic carbon and total nitrogen in relation to tillage, stubble management, and rotation," Australian Journal of Experimental Agriculture, vol. 35, no. 7, pp. 877–884, 1995.
33. G. Uhlen, "Long-term effects of fertilizers, manure, straw and crop rotation on total-N and total-C in soil," Acta Agriculturae Scandinavica, Section B, vol. 41, pp. 119–127, 1991.
34. H. Steinshamn, E. Thuen, M. A. Bleken, U. T. Brenoe, G. Ekerholt, and C. Yri, "Utilization of nitrogen (N) and phosphorus (P) in an organic dairy farming system in Norway," Agriculture, Ecosystems and Environment, vol. 104, no. 3, pp. 509–522, 2004.
35. E. Syväsalo, K. Regina, E. Turtola, R. Lemola, and M. Esala, "Fluxes of nitrous oxide and methane, and nitrogen leaching from organically and conventionally cultivated sandy soil in western Finland,"Agriculture, Ecosystems and Environment, vol. 113, no. 1–4, pp. 342–348, 2006.
36. D. J. Greenwood, F. Gastal, G. Lemaire, A. Draycott, P. Millard, and J. J. Neeteson, "Growth rate and % N of field grown crops: theory and experiments," Annals of Botany, vol. 67, no. 2, pp. 181–190, 1991.
37. F. E. Khasawneh and E. C. Doll, "The use of phosphate rock for direct application to soils," Advances in Agronomy, vol. 30, no. C, pp. 159–206, 1979.

38. M. Odlare, M. Pell, and K. Svensson, "Changes in soil chemical and microbiological properties during 4 years of application of various organic residues," Waste Management, vol. 28, no. 7, pp. 1246–1253, 2008.
39. H. Bengtsson, I. Öborn, S. Jonsson, I. Nilsson, and A. Andersson, "Field balances of some mineral nutrients and trace elements in organic and conventional dairy farming—a case study at Örjebyn, Sweden," European Journal of Agronomy, vol. 20, no. 1-2, pp. 101–116, 2003.
40. J. Verloop, J. Oenema, S. L. G. Burgers, H. F. M. Aarts, and H. van Keulen, "P-equilibrium fertilization in an intensive dairy farming system: effects on soil-P status, crop yield and P leaching,"Nutrient Cycling in Agroecosystems, vol. 87, no. 3, pp. 369–382, 2010.
41. H. Riley, "Long-term fertilizer trials on loam soil at Møystad, south-eastern Norway: crop yields, nutrient balances and soil chemical analyses from 1983 to 2003," Acta Agriculturae Scandinavica Section B, vol. 57, no. 2, pp. 140–154, 2007.
42. L. Taiz and E. Zeiger, Plant Physiology, Sinauer Associates, Sunderland, Mass, USA, 3rd edition, 2002.
43. S. Mathews, J. P. I. Tritschler, and S. C. Miyasaka, "Phosphorus management and sustainability," inGrass for Daily Cattle, J. H. Cherney and D. J. R. Cherney, Eds., pp. 193–222, CABI International, Wallingford, UK, 1998.
44. S. D. Heming, "Potassium balances for arable soils in southern England 1986–1999," Soil Use and Management, vol. 20, no. 4, pp. 410–417, 2004.
45. L. Blake, S. Mercik, M. Koerschens et al., "Potassium content in soil, uptake in plants and the potassium balance in three European long-term field experiments," Plant and Soil, vol. 216, no. 1-2, pp. 1–14, 1999.
46. I. Öborn, Y. Andrist-Rangel, M. Askekaard, C. A. Grant, C. A. Watson, and A. C. Edwards, "Critical aspects of potassium management in agricultural systems," Soil Use and Management, vol. 21, pp. 102–112, 2005.
47. A. F. Øgaard and S. Hansen, "Potassium uptake and requirement in organic grassland farming,"Nutrient Cycling in Agroecosystems, vol. 87, no. 1, pp. 137–149, 2010.

48. Y. Andrist-Rangel, A. C. Edwards, S. Hillier, and I. Öborn, "Long-term K dynamics in organic and conventional mixed cropping systems as related to management and soil properties," Agriculture, Ecosystems and Environment, vol. 122, no. 4, pp. 413–426, 2007
49. J. Holmqvist, A. F. Øgaard, I. Öborn, A. C. Edwards, L. Mattsson, and H. Sverdrup, "Application of the Profile model to estimate potassium release from mineral weathering in Northern European agricultural soils," European Journal of Agronomy, vol. 20, no. 1-2, pp. 149–163, 2003
50. H. Kuhlmann, "Importance of the subsoil for the K nutrition of crops," Plant and Soil, vol. 127, no. 1, pp. 129–136, 1990.
51. L. E. De-Bashan and Y. Bashan, "Recent advances in removing phosphorus from wastewater and its future use as fertilizer (1997–2003)," Water Research, vol. 38, no. 19, pp. 4222–4246, 2004.
52. A.-G. Roer, A. Korsaeth, T. M. Henriksen, O. Michelsen, and A. H. Strømman, "The influence of system boundaries on life cycle assessment of grain production in central southeast Norway,"Agricultural Systems, vol. 111, pp. 75–84, 2012

Chapter 4

The Extent of Grain Yield and Plant Growth Enhancement by Plant Growth-Promoting Broad-Spectrum Streptomyces Sp. in Chickpea

Subramaniam Gopalakrishnan, Vadlamudi Srinivas, Gottumukkala Alekhya, Bandikinda Prakash, Himabindu Kudapa, Abhishek Rathore, and Rajeev Kumar Varshney

International Crops Research Institute for the Semi-Arid Tropics (ICRISAT), Patancheru 502 324, Telangana, India

ABSTRACT

The physiological and molecular responses of five strains of *Streptomyces* sp. (CAI-17, CAI-68, CAI-78, KAI-26 and KAI-27), with their proven potential for charcoal rot disease control in sorghum and plant growth-promotion (PGP) in sorghum and rice, were studied to understand the mechanisms causing the beneficial effects. In this investigation, those five strains were evaluated for their PGP capabilities in chickpea in the 2012–13 and 2013–14 post-rainy seasons. All of the *Streptomyces* sp. strains exhibited enhanced nodule number, nodule weight, root weight and shoot weight at 30 days after sowing (DAS) and pod number, pod weight, leaf area, leaf weight and stem weight at 60 DAS in both seasons over the un-inoculated control. At crop maturity, the *Streptomyces* strains had enhanced stover yield, grain yield, total dry matter and seed number plant^{-1} in both seasons over the un-inoculated control. In the rhizosphere, the *Streptomyces* sp. also significantly enhanced microbial biomass carbon, dehydrogenase activity, total nitrogen, available phosphorous and organic carbon in both seasons over the un-inoculated control. Of the five strains of *Streptomyces* sp., CAI-17, CAI-68 and CAI-78 were superior to KAI-26 and KAI-27 in terms of their effects on root and shoot development, nodule formation and crop productivity. Scanning electron microscopy (SEM) micrographs had revealed the success in colonization of the chickpea roots by all five strains. Quantitative real-time PCR (qRT-PCR) analysis of selected PGP genes of actinomycetes revealed the selective up-regulation of indole-3-acetic acid (IAA)-related and siderophore-related genes by CAI-68 and of -1,3-glucanase genes by KAI-26.

INTRODUCTION

In recent years, the use of plant growth-promoting (PGP) bacteria has become one of the most attractive options for enhancing the sustainability of agricultural systems in many parts of the world due to their eco-friendliness, low production cost and reduced use of non-renewable resources. PGP bacteria promote plant growth through various mechanisms such as nitrogen fixation, the solubilization of phosphorus, the chelation of iron, the secretion of plant growth hormones and the inhibition of phytopathogens. Bacteria such as

Bacillus sp., *Pseudomonas* sp. and *Streptomyces* sp. were found to play a major role both in mobilizing and acquiring nutrients and in controlling phytopathogens (Postma et al. [2003]; Perner et al. [2006]; Gopalakrishnan et al. [2011a], [b], [c]; Jalilian et al. [2012]). The presence of PGP bacteria in the rhizosphere has been shown to enhance shoot growth, root growth, root hair development, plant hormone concentrations, nitrogen fixation, the solubilization of minerals and the suppression of pathogens (Shaukat et al. [2006]; Lucas et al. [2009]; Richardson et al. [2009]; Gopalakrishnan et al. [2012a], [b]). The PGP potential of *Streptomyces* sp. has been demonstrated on tomato, wheat, rice, bean and pea (Tokala et al. [2002]; Nassar et al. [2003]; El-Tarabily [2008]; Sadeghi et al. [2012]; Gopalakrishnan et al. [2013], [2014a]).*Streptomyces* sp. are known to promote plant growth either by producing indole-3-acetic acid (IAA) and siderophores and/or by inhibiting soil-borne fungal pathogens (Aldesuquy et al.[1998]; Trejo-Estrada et al. [1998]; Tokala et al. [2002]; Macagnan et al. [2008]; Gopalakrishnan et al. [2011a], [b], [2013]; [2014a]).

Previously, we have demonstrated the potential of a set of eight *Streptomyces* strains (CAI-21, CAI-26, MMA-32, CAI-17, CAI-68, CAI-78, KAI-26 and KAI-27) isolated from herbal vermicompost, for the bio-control of charcoal rot disease, caused by *Macrophomina phaseolina*(Tassi) Goid., in sorghum (Gopalakrishnan et al. [2011b]) and for PGP in rice (Gopalakrishnan et al. [2012a], [2013]). The objectives of this investigation were to further evaluate five of the eight *Streptomyces* strains (CAI-17, CAI-68, CAI-78, KAI-26 and KAI-27) for their PGP traits in chickpea under field conditions, to demonstrate gene expression profiles by qRT-PCR analysis, and to ensure colonizing ability in chickpea by scanning electron microscopy (SEM) analysis.

MATERIALS AND METHODS

Streptomyces Strains

Five strains of *Streptomyces* spp., CAI-17 (GenBank accession number: JQ682619), CAI-68 (GenBank accession number: JQ682622), CAI-78 (GenBank accession number: JQ682623), KAI-26 (GenBank accession number: JQ682624) and KAI-27 (GenBank accession number:

JQ682625), previously reported by us to have capacity for the bio-control of charcoal rot in sorghum (Gopalakrishnan et al. [2011b]) and PGP in rice (Gopalakrishnan et al. [2013]), were further studied in the present investigation.

Evaluation of Streptomyces sp. for PGP Potential on Chickpea under Field Conditions

This experiment was carried out during the 2012–13 and 2013–14 post-rainy cropping seasons at ICRISAT, Patancheru (17°30′ N; 78°16′ E; altitude 549 m) in peninsular India. Soils at the experimental site are classified as Vertisols (fine montmorillonitic isohyperthermic typic pallustert) having 52% clay, 21% silt and 26% sand, with an alkaline pH of 7.7 – 8.5 and an organic carbon content of 0.4 – 0.6%. The soil depth of the field used was ≥ 1.2 m, and this soil retained approximately 200 mm of plant-available water in a 120-cm (maximum rooting depth by chickpea) soil profile. The mineral content of the top 15 cm of rhizosphere soil includes 24.7 mg kg^{-1} soil of available nitrogen, 8.6 mg kg^{-1} soil of available phosphorous and 298 mg kg^{-1} soil of available potassium. The field was kept fallow except for this post-rainy season crop. The fields were prepared into broad beds and furrows with beds 1.2 m wide flanked by 0.3-m furrows in both seasons. Surface application and incorporation of 18 kg N ha^{-1} and 20 kg P ha^{-1} as di-ammonium phosphate (DAP) were performed in both seasons. The experiment was laid out in a randomized complete block design (RCBD) with three replicates and plot sizes of 4 m × 3 ridges (rows).

The five strains of *Streptomyces* spp. were cultured individually on a starch casein broth at 28°C for five days. Seeds of chickpea variety ICCV 2 (with a grain yield potential of 1.1 to 1.2 t ha^{-1}) were treated with a *Streptomyces* sp. (containing 10^8 CFU ml^{-1}) for 50 min and sown by hand on 11 October 2012 in the first year and 1 November 2013 in the second year in rows 30 cm apart at a depth of 4–5 cm to achieve an estimated plant stand density of at least 26 plants m^{-2}. *Streptomyces* sp. (1000 ml; 10^8 CFU mL^{-1}) was applied once every 15 days on the soil close to the plant until the flowering stage. Control plots were maintained without the application of *Streptomyces* strains. The plots were irrigated on the 21st and 49th days after sowing. The crop was kept weed-free by manual weeding. No serious insect pest or phytopathogen attacks were observed during the cropping period. The crop was harvested manually on 18 January 2013 in the first year and 6 Feb 2014 in the second year. In both the 2012–13 and 2013–14 seasons, the nodule number (plant^{-1}), nodule weight (mg plant^{-1}), root weight (mg plant^{-1}) and shoot weight (g plant^{-1}) were recorded at 30 days after sowing (DAS), and at 60 DAS, the plant height (cm), pod number (plant^{-1}), pod weight (g plant^{-1}), leaf area (cm^{-2} plant^{-1}) and stem weight (g plant^{-1}) were recorded. At both harvests, stover yield (t ha^{-1}), grain yield (t ha^{-1}), total dry matter (t ha^{-1}), 1000-seed weight (g), pod weight (g plant^{-1}), seed number (plant^{-1}) and seed weight (g plant^{-1}) were recorded. Soil samples were collected from the top 15 cm of the soil profile at the harvest stage and analysed for soil chemistry (total nitrogen [ppm], available phosphorous [ppm] and % organic carbon as per the protocols of Novozamsky et al. [1983], Olsen and Sommers [1982] and Nelson and Sommers[1982], respectively) and microbial biomass carbon (μg g^{-1} soil) by the fumigation method and dehydrogenase activity (μg TPF g^{-1} soil 24 h^{-1}) by the triphenyl formazan production method as per the protocols of Anderson and Domsch [1989] and Casida [1977], respectively.

Colonization of Streptomyces sp.

The roots of chickpea were examined for colonization by *Streptomyces* sp. by SEM (Bozzola and Russell [1998]). For this procedure, the seeds of chickpea variety ICCV 2 were surface-sterilized first with 2.5% sodium hypochlorite solution (for five minutes), followed by 70% ethanol (for five minutes), and then rinsed with sterilized water (eight times) before being allowed to sprout in a Petri plate overnight. The sprouted seeds were transferred into test*Streptomyces* strains (CAI-17, CAI-68, CAI-78, KAI-26 and KAI-27; grown in Bennett's broth separately) for an hour before being sown in pots containing sterilized coarse sand (six seeds/8" plastic pot). The soil was drenched with a booster dose of *Streptomyces* strains (5 ml per seedling; 10^8 CFU ml^{-1}) one week after sowing in sand. The pots were incubated for two weeks in a greenhouse at a temperature of 24 ± 2°C. At the end of the incubation period, chickpea seedlings were removed from the sand pots and the roots were washed in phosphate buffer (0.1 M; pH 7.2). The tip of the roots were cut into pieces 5 mm long and fixed in 2.5% glutaraldehyde in phosphate buffer for 24 h at 4°C. At the end of incubation, the samples were washed with phosphate buffer, postfixed in 2% osmium tetroxide for 4 h and dehydrated using a graded series of ethanol. The dehydrated samples were dried with critical-point liquid carbon dioxide as a transition fluid. The dried materials were adhered onto aluminium specimen mounts with double-stick adhesive tape. The samples were later coated with gold-palladium in an automated sputter coater (JEOL JFC-1600) and examined with a scanning electron microscope (JOEL-JSM 5600) as per the standardized protocols at RUSKA lab, College of Veterinary Science, Rajendranagar, Hyderabad, India. Observations of the presence of*Streptomyces* spores on root surfaces were recorded.

Gene Expression Profile

RNA Extraction

The five selected *Streptomyces* sp. (CAI-17, CAI-68, CAI-78, KAI-26 and KAI-27) were grown in Bennett's broth at 28°C for 72 h. Total RNA was extracted from all of the *Streptomyces*strains using the "NucleoSpin® RNA Plant" kit (Macherey-Nagel). The quality and quantity of RNA was estimated by a Nanodrop (Thermo Scientific) and RNA integrity was determined using a 2100 Bioanalyzer (Agilent).

Quantitative Real Time- PCR (qRT-PCR)

qRT-PCR was performed using the Applied Biosystems 7500 Real-Time PCR System with the SYBR green chemistry (Applied Biosystems, USA) as per the manufacturer's instructions. Well-characterized genes Spaepen et al. [2007] relating to IAA production (F: GTCACCGGGATCTTCTTCAAC; R: GATGTCGGTGTTCTTGTCCAG) and siderophore (F: ATCCTCAACACCCTGGTCTG; R: TCCTTGTACTGGTACGGGACTT) were collected from the UniprotKB database (http://www.uniprot.org/uniprot/) as described by Gopalakrishnan et al. ([2014a]). Similarly, genes for -1,3-glucanase (F: CCGAACACCACCTACTCCAC; R: CCAGGTTGAGGATCAGGAAG) were also selected for the study. Gene-specific primers for real-time PCR were designed using Primer 3 software (Rosen and Skaletsky [2000]). RNA polymerase principal sigma factor *HrdB* (SCO5820) (F: GGTCGAGGTCATCAACAAGC; R:

CTCGATGAGGTCACCGAACT) was used as the endogenous control. qRT-PCR reactions were performed as described earlier (Gopalakrishnan et al.[2014a]). The data obtained were analysed using the mean of the CT values of the three biological replicates that were normalized to the mean CT values of the endogenous gene. The expression ratios were calculated using the $2^{-\Delta\Delta Ct}$ method. Relative transcription levels are presented graphically.

Statistical Analysis

Analysis of variance) was performed by the GLM (General Linear Model) procedure of SAS software package (SAS Institute Inc [2013]). Isolate were tested for significance and their means with control were compared by using Dunnett's *t*-test.

RESULTS

As all of the strains of *Streptomyces* spp. (CAI-17, CAI-68, CAI-78, KAI-26 and KAI-27) enhanced different plant component performances, the means of all the five strains are presented. The *Streptomyces*-treated plots, by a mean of the five strains, enhanced agronomic performance in terms of nodule numbers (up to 61%), nodule weight (up to 44%) and shoot weight (up to 28%) at 30 DAS and plant height (up to 5%), pod numbers (up to 79%), pod weight (up to 89%), leaf area (up to 29%), leaf weight (up to 42%) and stem weight (up to 6%) at 60 DAS in both seasons (2012–13 and 2013–2014) over the un-inoculated control plots (Tables 1 and 2). At chickpea crop maturity, the *Streptomyces*-treated plots also enhanced stover yield (up to 41%), grain yield (up to 16%), total dry matter (up to 26%), pod weight (up to 7%), seed number (up to 21%) and seed weight (up to 11%) in both seasons over the un-inoculated control plots (Table 3). At crop maturity, in the top 15-cm of rhizosphere soil, the*Streptomyces* sp. treatments enhanced the microbial biomass carbon (up to 59%) and dehydrogenase activity (up to 21%), as well as soil mineral nutrient contents, including total N (up to 8%), available P (up to 45%) and % organic carbon (up to 7%), in both seasons over the un-inoculated control plots (Tables 4 and 5).

Table 1: Response of various plant components of chickpea to the application of different strains of*Streptomyces*sp. at 30 days after sowing in a Vertisol during 2012–13 and 2013–14 post-rainy seasons

	Year 2012–13				**Year 2013–14**			
Isolate	**Nodule number ($plant^{-1}$)**	**Nodule weight (mg $plant^{-1}$)**	**Root weight (mg $plant^{-1}$)**	**Shoot weight (g $plant^{-1}$)**	**Nodule number ($plant^{-1}$)**	**Nodule weight (mg $plant^{-1}$)**	**Root weight (mg $plant^{-1}$)**	**Shoot weight (g $plant^{-1}$)**
CAI-17	20*	36	186*	2.11*	51	228	192	1.76
CAI-68	19*	37	173	1.76*	60*	240*	205*	1.91*
CAI-78	21*	71*	184	1.81*	59*	242*	207*	1.87
KAI-26	20*	29	172	1.39	50	228	183	1.75
KAI-27	19*	36	172	1.55	50	224	186	1.74
Control	12	29	171	1.35	49	221	168	1.72
Mean	19	40	176	1.66	53	231	190	1.79
SEM±	1.04***	4.1***	3.3*	0.06***	0.8***	4.2*	5.75***	0.04*
CV%	10	18	3	6	3	3	5	4

SEM = Standard Error of Mean; CV = Coefficients of variation; * = significant at $p < 0.05$, *** = significant at $p < 0.001$; According to Dunnett's *t*-test, Isolate mean values are compared with control at $p < 0.05$.

Gopalakrishnan *et al.*

Gopalakrishnan *et al. SpringerPlus* 2015 4:31, mdoi:10.1186/s40064-015-0811-3

Table 2: Response of various plant components of chickpea to the application of different strains of *Streptomyces* sp. at 60 days after sowing in a Vertisol during 2012–13 and 2013–14 post-rainy seasons

	Year 2012–13						Year 2013–14					
Isolate	Plant height (cm)	Pod number (plant^{-1})	Pod weight (g plant^{-1})	Leaf area (cm^{-2}plant^{-1})	Leaf weight (g plant^{-1})	Stem weight (g plant^{-1})	Plant height (cm)	Pod number (plant^{-1})	Pod weight (g plant^{-1})	Leaf area (cm^{-2}plant^{-1})	Leaf weight (g plant^{-1})	Stem weight (g plant^{-1})
CAI-17	54*	65*	3.26*	1042*	7.25*	6.55*	49*	58	4.18	781*	4.45	4.17
CAI-68	57*	72*	4.47*	868*	5.30*	4.87*	48	66*	4.32	667	4.05	4.75*
CAI-78	51	91*	3.41*	770*	6.39*	4.92*	50*	87*	5.58*	878*	4.67	4.43
KAI-26	51	72*	3.52*	906*	6.38*	5.18*	49*	58	4.18	832*	4.54	4.17
KAI-27	48	83*	4.50*	739	5.56*	4.75*	47	56	4.49*	633	4.00	4.18
Control	50	43	2.04	670	4.36	3.36	47	57	4.14	626	3.71	4.11
Mean	52	71	3.53	833	5.87	4.94	48	64	4.48	736	4.24	4.30
SE±	0.4***	1.9***	0.151***	31.0***	0.212***	0.071***	0.6*	3.3***	0.079***	15.1***	0.199*	0.096**
CV%	1	5	7	6	6	3	2	9	3	4	8	4

SEM = Standard Error of Mean; CV = Coefficients of variation; * = significant at $p < 0.05$, ** = significant at $p < 0.01$, *** = significant at $p < 0.001$; According to Dunnett's *t*-test, isolate mean values re compared with control at $p < 0.05$.

Gopalakrishnan *et al.*

Gopalakrishnan *et al.* *SpringerPlus* 2015 4:31, doi:10.1186/s40064-015-0811-3

Table 3: Response of grain yield and its components to the application of different strains of*Streptomyces*sp. to chickpea in a Vertisol during 2012–13 and 2013–14 post-rainy

	Year 2012–13							Year 2013–14						
Isolate	Stover yield (t ha^{-1})	Grain yield (t ha^{-1})	**TDM (t ha^{-1})**	**1000-seed weight (g)**	**Pod weight (g plant^{-1})**	**Seed number (plant^{-1})**	**Seed weight (g plant^{-1})**	**Stover yield (t ha^{-1})**	**Grain yield (t ha^{-1})**	**TDM (t ha^{-1})**	**1000-seed weight (g)**	**Pod weight (g plant^{-1})**	**Seed number (plant^{-1})**	**Seed weight (g plant^{-1})**
CAI-17	1.42*	2.01*	3.43*	220	22.2*	72*	16.7	1.75	1.70*	3.45	196	18.0*	72	15.0*
CAI-68	1.31*	1.89*	3.20*	221	18.2	72*	15.4	2.03	1.70*	3.72*	198*	17.3	70	15.3*
CAI-78	1.91*	2.16*	4.08*	227*	18.2	55	14.1	2.19*	1.84*	4.03*	200*	21.0*	82	15.9*
KAI-26	1.33*	1.96*	3.28*	231*	19.8*	63*	14.6	1.69	1.76*	3.45	196	17.4	65*	14.3
KAI-27	1.81*	1.81*	3.62*	222	17.2	63*	15.4	1.82	1.68*	3.51	196	17.5	67	13.6
Control	1.11	1.69	2.80	220	18.0	54	13.8	1.68	1.55	3.23	195	17.0	66	13.3
Mean	1.48	1.92	3.40	223	18.9	63	15.0	1.86	1.71	3.57	197	18.0	70	14.6
SEM±	0.020***	0.022***	0.034***	1.4***	0.24***	0.5***	0.42**	0.093*	0.019***	0.098**	0.7**	0.23***	3.3*	0.30***
CV%	2	2	2	1	2	2	5	9	2	5	1	2	8	4

TDM = total dry matter; SEM = Standard Error of Mean; CV% = Coefficients of variation; * = significant at $p < 0.05$, ** = significant at $p < 0.01$, *** = significant at $p < 0.001$; According to Dunnett's *t*-test, isolate mean valus are compared with control at $p < 0.05$. Gopalakrishnan *et al.*

Gopalakrishnan *et al. SpringerPlus* 2015 4:31, doi:10.1186/s40064-015-0811-3

seasons

Table 4: Response of microbial biomass carbon and dehydrogenase activity in the chickpea rhizosphere to the application of different strains ofStrepto-mycesspp. in a Vertisol during 2012–13 and 2013–14 post-rainy seasons

	Year 2012–13		Year 2013–14	
Isolate	**Microbial biomass C (µg g^{-1} soil)**	**Dehydrogenase activity (µg TPF g^{-1} soil 24 h^{-1})**	**Microbial biomass C (µg g^{-1} soil)**	**Dehydrogenase activity (µg TPF g^{-1} soil 24 h^{-1})**
CAI-17	1221*	52.7	974*	68.6*
CAI-68	1295*	56.8	1000*	71.5*
CAI-78	1424*	63.6*	908*	67.8
KAI-26	893*	53.0	842*	68.8*
KAI-27	1132*	56.8	444	59.0
Control	752	51.9	550	55.7
Mean	1119	55.8	786	65.2
SEM (±)	6.7***	1.74**	47.1**	2.41*
CV%	1	4	9	5

SEM = Standard Error of Mean; CV = Coefficients of variation; * = significant at $p < 0.05$, ** = significant at $p < 0.01$, *** = significant at $p < 0.001$; According to Dunnett's *t*-test, isolate mean values are compared with control at $p < 0.05$.

Gopalakrishnan *et al.*

Gopalakrishnan *et al.* *SpringerPlus* 2015 4:31, doi:10.1186/s40064-015-0811-3

Table 5: Response of rhizosphere soil chemical properties of chickpea at its maturity to the application of different strains of*Streptomyces*sp. in a Vertisol during the 2012–13 and 2013–14 post-rainy seasons

	Year 2012–13			**Year 2013–14**		
Isolate	**Total N (ppm)**	**Available P (ppm)**	**Organic carbon (%)**	**Total N (ppm)**	**Available P (ppm)**	**Organic carbon (%)**
CAI-17	656	11.2*	0.51*	737	11.1*	0.56
CAI-68	703*	10.4	0.49*	744	10.2	0.54
CAI-78	714*	12.2*	0.56*	815*	11.3*	0.67*
KAI-26	666	10.8*	0.45	724	10.2	0.57
KAI-27	668	10.8*	0.47	720	6.9	0.58
Control	632	10.1	0.47	735	6.9	0.55
Mean	673	10.9	0.49	746	9.4	0.58
SEM±	13.6*	0.13***	0.004***	15.2*	0.67**	0.016*
CV%	3	2	1	3	10	4

\SEM = Standard Error of Mean; CV = Coefficients of variation; * = significant at $p < 0.05$, ** = significant at $p < 0.01$, *** = significant at $p < 0.001$; According to Dunnett's *t*-test, isolate mean values are compared with control at $p < 0.05$.

Gopalakrishnan *et al.*

Gopalakrishnan *et al. SpringerPlus* 2015 4:31, doi:10.1186/s40064-015-0811-3

SEM analysis of chickpea roots revealed a remarkable degree of colonization by *Streptomyces*sp. Roots from inoculated plants exhibited significant surface colonization by *Streptomyces* sp., while those from un-inoculated plants did not. The sporulation of *Streptomyces* sp. on the surface cell layer of chickpea roots was clearly evident for all five strains. The hyphae of*Streptomyces* sp. were also found to penetrate the surface cell layer of chickpea roots (Figure 1).

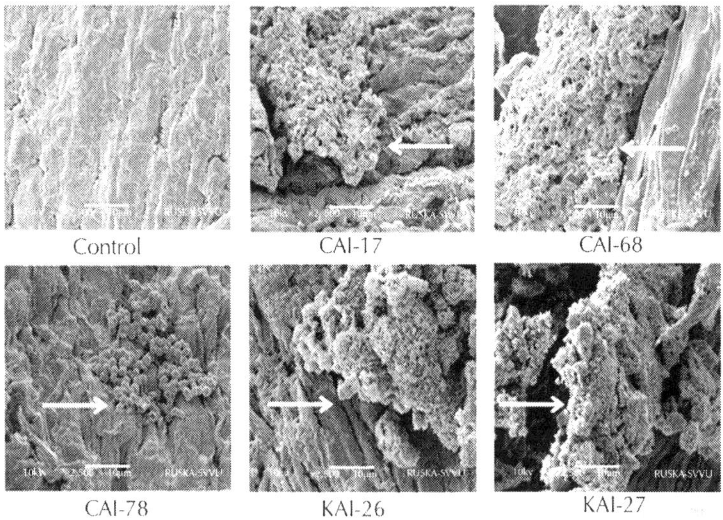

Figure 1: Scanning electron microscopy photographs of the five*Streptomyces*sp. showing colonization on the roots of chickpea. Note: Arrows indicate chickpea roots colonized by PGP Streptomyces strains CAI-17, CAI-68, CAI-78, KAI-26, and KAI-27.

Gene expression profiles of plant-growth-promoting genes, IAA, siderophore and -1,3-glucanase revealed overall up-regulation in the *Streptomyces* strains studied. The gene IAA showed higher up-regulation in the strain CAI-68 (13.7-fold), while no significant variation in the expression pattern was observed in the other four strains. The siderophore gene was highly up-regulated in strain CAI-68 (10.8-fold), followed by KAI-27 (3.6), and -1,3-glucanase expression was up-regulated in KAI-26 (9.3-fold) (Figure 2).

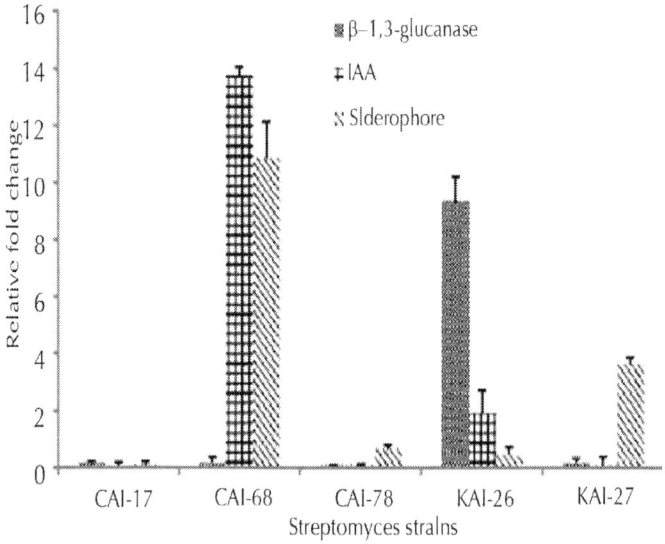

Figure 2: Expression profiling of PGP genes of*Streptomyces*sp. through quantitative real-time PCR assays.

DISCUSSION

The five strains of *Streptomyces* spp. (CAI-17, CAI-68, CAI-78, KAI-26 and KAI-27), which are known for biocontrol of charcoal rot in sorghum and PGP in sorghum (Gopalakrishnan et al.[2011b]) and rice (Gopalakrishnan et al. [2013]), were evaluated for their PGP effects in chickpea under field conditions. In the present study, nodules were observed in the roots of chickpea on both the seasons (2012–13 and 2013–2014) though rhizobia were neither inoculated on seed nor applied on soil indicating the presence of native rhizobia in the soil. Rhizobia were always noted in theVertisols of ICRISAT (Gopalakrishnan et al. [2014b]). The increase in the number of nodules per plant in 2013–14 over 2012–13 could be due to the build-up of native rhizobia specific to chickpea during the cropping season of 2012–13 and hence subsequently observed in the following season, 2013–14. Of the five *Streptomyces* sp. applied, CAI-78 significantly enhanced all of the agronomic characteristics and yield components the most, including the nodule numbers and nodule weight. The enhancement effects of the other strains were, in descending order, CAI-68, CAI-17, KAI-26 and KAI-27.

The highest magnitude of enhanced soil biological activities and mineral nutrient properties was found with CAI-78 treatment, followed by CAI-68, CAI-17, KAI-26 and KAI-27. The enzyme dehydrogenase is regarded as an indicator of total life in the soil and a strong indicator of biological activity. Similarly, increased microbial biomass C and dehydrogenase activity have been reported in fields upon the use of biological options for soil fertilization and insect-pests control (Carpenter-Boggs et al. [2000]; Sharma and Singh [2004]). Superior soil nutrient status (N, P, K and % organic carbon) has been reported on organic farms compared to soils treated with chemical pesticides and fertilizers (Sharma and Singh [2004]; Singh et al. [2004]). In the present study, the enhancement of soil biological and mineral nutrient properties with*Streptomyces* treated plots over un-inoculated control could be due to the presence of more microbial activity in the rhizosphere which leads to beneficial functions for crops such as plant growth-promotion, nitrogen fixation, phosphate solubilization and protection against pathogens.

The PGP influence of *Streptomyces* sp. had previously been reported in various agriculturally important crops (Tokala et al. [2002]; Nassar et al. [2003]; El-Tarabily [2008]; Sadeghi et al.[2012]; Gopalakrishnan et al. [2014a]). PGP agents having a broad spectrum of activities offer effective and novel strategies not only for the improvement of crop growth and yield but also for the control of insect pests and pathogens that affect crop plants. In addition to suppressing plant pathogens and insect pests by secreting secondary metabolites, such as antibiotics, some PGP agents also elicit induced systemic resistance against a broad range of insects and pathogens (Jetiyanon and Kloepper [2002]; Ryu et al. [2007]). Although the *Streptomyces* spp. used in this study excrete antimicrobial compounds, antagonism tests by the poison food technique indicate that none of the strains inhibit the growth of nodulating bacteria *Rhizobium leguminosarum* on yeast extract mannitol agar media (data not shown). Thus, the*Streptomyces* spp., appears to be compatible with *Rhizobium*.

Colonization of chickpea roots by *Streptomyces* spp., at the right time and place is essential for enhanced PGP activity. Successful host-microbe interaction is essential, which depends on the presence of a sufficient population of bacteria, as well as the rhizosphere competence, root-colonizing ability and PGP capability of the bacteria (Lugtenberg and Dekkers [1999]). The data for grain and stover yield, root mass and other agronomical traits, the chemical and biological activities of the rhizosphere soil along with the SEM micrograph clearly establish that the PGP effects of the *Streptomyces* spp., had been caused by successful colonization of the inoculated chickpea roots.

The mechanism by which the five *Streptomyces* sp. could consistently enhance agronomical and yield traits on sorghum, rice (from our previous study) and chickpea can be attributed to their enzymatic activities, such as the ability to produce siderophores and indole acetic acid and -1,3-glucanase (Gopalakrishnan et al. [2011b]; [2013]). Siderophores are known to form stable organic complexes with heavy metals such as Cu, Cd, In, U, Np, Al, Pb, Zn and Ga and increase the soluble metal concentration (Rajkumar et al. [2010]) in the soil system, thus alleviating the various heavy metal stresses imposed on plants. Siderophores also act as solubilizing agents of iron from minerals when iron nutrition becomes a limitation (Indiragandhi et al. [2008]). IAA-producing microorganisms are known to stimulate seed germination, initiate adventitious and lateral root formation and increase root length and surface area, thereby providing the host plant greater access to soil nutrients and water (Ahemad and Kibret [2014]). The cell wall of plant pathogens is composed of layers of -1,3-glucan (as in the case of*Fusarium oxysporum*, the causal agent of wilt disease in many crops) and lysis of these layers by -1,3-glucanase-producing microorganisms leads to the leakage of cell contents and the collapse of the pathogenic fungi (Singh et al. [1999]).

Our previous results documented the production of IAA (0.22 – 0.95%), -1,3 glucanase (0.2 – 2.92 units) and siderophore (qualitative; 1–3 rating scale) by all of these *Streptomyces* strains (Gopalakrishnan et al. [2011b]; Gopalakrishnan et al. [2013]). Further characterization on gene expression profiles of these PGP genes revealed overall up-regulation in all the *Streptomyces*strains. However, no significant change in expression profiles of all three genes, IAA, siderophore and -1,3-glucanase, was observed in isolates CAI-17 and CAI-78. The relative expression profiles of these genes in CAI-17 and CAI-78 do not correlate with the PGP capabilities observed under field conditions. This may be due to analysis of the gene expression profiles in only one biological replication of the strains. The exact role of these genes is to be precisely elucidated with more biological replications. However, validation of the IAA, siderophore and -1,3-glucanase genes confirmed the insights gained from *in vitro* PGP attributes of the studied *Streptomyces* strains. Hence, it is concluded that the *Streptomyces*strains studied in this investigation contains multiple PGP traits and can therefore be exploited not only for PGP but also for biological control of plant pathogens.

The five *Streptomyces* spp. used in this study are apparently well adapted not only to the sorghum and rice rhizospheres, as reported earlier, but also to the chickpea rhizosphere, as demonstrated in the current investigation. Additionally, the five *Streptomyces* spp. contain a broad range of PGP abilities and demonstrate multiple beneficial actions. Therefore, the five*Streptomyces* spp. used in this study are potential candidates for the discovery of novel secondary metabolites and their usefulness in host plant resistance against a range of pathogens and insect pests. These *Streptomyces* sp. strains can be useful in furthering the use of eco-friendly bio-pesticides and bio-fertilizers in integrated pest, disease and nutrition management programs. However, there is a need to determine the effectiveness of these*Streptomyces* strains under multi-location trials and to understand the nature of their interaction with other native soil micro-flora and micro-fauna with host plants and the environment.

AUTHORS' CONTRIBUTIONS

SG, VS, GA, BH, HK and RKV carried out the lab and field studies. AR done the statistical analysis and SG drafted the manuscript. All authors read and approved the final manuscript.

ACKNOWLEDGEMENTS

This work has been undertaken as part of the CGIAR Research Program on Grain Legumes. We thank Dr. M Lakshman, Associate Professor, Ruska Lab, College of Veterinary Science, Rajendranagar, Hyderabad, for SEM analysis and all of the staff of the biocontrol unit of ICRISAT including M/s PVS Prasad, P Manohar, B Nagappa, D Barath, A Jabbar and S Rohini for their significant contributions in the laboratory and field studies. We also thank Dr L Krishnamurthy, Consultant, Grain Legumes, ICRISAT for his critical review and useful suggestions for improving this manuscript.

REFERENCES

1. Ahemad M, Kibret M (2014) Mechanisms and applications of plant growth promoting rhizobacteria: Current perspective. J King Saud University 26:1-20 Anderson TH, Domsch KH (1989) Ratios of microbial biomass carbon to total organic carbon in arable soils. Soil Biol Biochem 21:471-479
2. Aldesuquy I IS, Mansour FA, Abo-Hamed SA (1998) Effect of the culture filtrates of*Streptomyces* on growth and productivity of wheat plants. Folia Microbiol 43:465-470
3. Bozzola JJ, Russell LD (1998) "Specimen Preparation for Scanning Electron Microscopy". 3rd Chapter. In: Electron Microscopy Principals and Techniques for Biologists. Jones and Barlett publishers, Sudbury, Massachusetts. pp 19-24 54–55, 63–67
4. Carpenter-Boggs L, Kennedy AC, Reganold JP (2000) Organic and biodynamic management: Effects on soil biology. Soil Sci Soc Am J 64:1651-1659
5. Casida LE (1977) Microbial metabolic activity in soil as measured by dehydrogenase determinations. Appl Environ Microbiol 34:630-636
6. El-Tarabily KA (2008) Promotion of tomato (*Lycopersicon esculentum* Mill.) plant growth by rhizosphere competent 1-aminocyclopropane-1-carboxylic acid deaminase–producing*Streptomycete* actinomycetes. Plant Soil 308:161-174
7. Gopalakrishnan S, Humayun P, Kiran BK, Kannan IGK, Vidya MS, Deepthi K, Rupela O (2011) Evaluation of bacteria isolated from rice rhizosphere for biological control of charcoal rot of sorghum caused by *Macrophomina phaseolina* (Tassi) Goid. World J Microbiol Biotechnol 27:1313-1321
8. Gopalakrishnan S, Humayun P, Vadlamudi S, Vijayabharathi R, Bhimineni RK, Rupela O (2012) Plant growth-promoting traits of *Streptomyces* with biocontrol potential isolated from herbal vermicompost. Biocontrol Sci Technol 22:1199-1210

9. Gopalakrishnan S, Kiran BK, Humayun P, Vidya MS, Deepthi K, Rupela O (2011) Biocontrol of charcoal-rot of sorghum by actinomycetes isolated from herbal vermicompost. Afr J Biotechnol 10:18142-18152
10. Gopalakrishnan S, Pande S, Sharma M, Humayun P, Kiran BK, Sandeep D, Vidya MS, Deepthi K, Rupela O (2011) Evaluation of actinomycete isolates obtained from herbal vermicompost for biological control of *Fusarium* wilt of chickpea. Crop Prot 30:1070-1078
11. Gopalakrishnan S, Vadlamudi S, Apparla S, Bandikinda P, Vijayabharathi R, Bhimineni RK, Rupela O (2013) Evaluation of *Streptomyces* spp. for their plant growth-promotion traits in rice. Can J Microbiol 59:534-539
12. Gopalakrishnan S, Vadlamudi S, Bandikinda P, Sathya A, Vijayabharathi R (2014b) Plant growth-promoting traits of Pseudomonas geniculata isolated from chickpea nodules. 3Biotech. DOI 10.1007/s13205-014-0263-4
13. Gopalakrishnan S, Vadlamudi S, Bandikinda P, Sathya A, Vijayabharathi R, Rupela O, Kudapa H, Katta K, Varshney RK (2014) Evaluation of *Streptomyces* strains isolated from herbal vermicompost for their plant growth-promotion traits in rice. Microbiol Res 169:40-48
14. Gopalakrishnan S, Upadhyaya HD, Vadlamudi S, Humayun P, Vidya MS, Alekhya G, Singh A, Vijayabharathi R, Bhimineni RK, Seema M, Rathore A, Rupela O (2012) Plant growth-promoting traits of biocontrol potential bacteria isolated from rice rhizosphere. SpringerPlus 1:71
15. Indiragandhi P, Anandham R, Madhaiyan M, Sa TM (2008) Characterization of plant growth-promoting traits of bacteria isolated from larval guts of diamondback moth*Plutella xylostella* (Lepidoptera; Plutellidae). Curr Microbiol 56:327-333
16. Jalilian J, Modarrcs-Sanavy SAM, Saberali SF, Sadat-Asilan K (2012) Effects of the combination of beneficial microbes and nitrogen on sunflower seed yields and seed quality traits under different irrigation regimes. Field Crop Res 127:26-
17. Jetiyanon K, Kloepper JW (2002) Mixtures of plant growth promoting rhizobacteria for induction of systemic resistance against multiple plant diseases. Biol Control 24:285-291

18. Lucas JA, Solano BR, Montes F, Ojeda J, Megias M, Manero FJG (2009) Use of two PGPR strains in the integrated management of blast disease in rice (*Oryza sativa*) in Southern Spain. Field Crop Res 114:404-410
19. Lugtenberg BJJ, Dekkers LC (1999) what makes *Pseudomonas* bacteria rhizosphere competent? Environ Microbiol 1:9-13
20. Macagnan D, Romeiro RDA, Pomella AMV, deSouza JT (2008) Production of lytic enzymes and siderophores, and inhibition of germination of basidiospores of*Moniliophthora* (ex *Crinipellis*) *perniciosa* by phylloplane actinomycetes. Biol Control 47:309-314
21. Nassar AH, El-Tarabily KA, Sivasithamparam K (2003) Growth promotion of bean (*Phaseolus vulgaris* L.) by a polyamine producing isolate of *Streptomyces griseoluteus*. Plant Growth Reg 40:97-106
22. Nelson DW, Sommers LE (1982) Total Organic Carbon and Organic matter. In: Page AL, Miller RH, Keeney DR (eds) Methods of Soil Analysis, Part 3, Chemical and Microbiological Properties, SSSA, Madison, WI. pp 39-579
23. Novozamsky I, Houba VJG, Van ECKR, vanVark W (1983) A novel digestion technique for multiple element analysis. Comm Soil Sci Plant Anal 14:239-249
24. Olsen SR, Sommers LE (1982) Phosphorus. In: Page AL (ed) Methods of Soil Analysis, Agron No 9, Part 2, 'Chemical and Microbial properties', 2nd edn. American Society of Agronomy, Madison WI, USA. pp 403-430
25. Perner H, Schwarz D, George E (2006) Effect of mycorrhizal inoculation and compost supply on growth and nutrient uptake of young leek plants grown on peat-based substrates. HORTSCI 41:628-632
26. Postma J, Montanari M, Van den Boogert PHJF (2003) Microbial enrichment to enhance disease suppressive activity of compost. Eur J Soil Biol 39:157-163
27. Rajkumar M, Ae N, Prasad MNV, Freitas H (2010) Potential of siderophore-producing bacteria for improving heavy metal phytoextraction. Trends Biotechnol 28:142-149

28. Richardson AE, Barea JM, Mcneill AM, Combaret CP (2009) Acquisition of phosphorus and nitrogen in the rhizosphere and plant growth promotion by microorganisms. Plant Soil 321:305-339
29. Rosen S, Skaletsky HJ (2000) Primer 3 on the WWW for General Users and for Biologist Programmers. In: Krawetz S, Misener S (eds) Bioinformatics Methods and Protocols: Methods in Molecular Biology, Humana Press, Totowa, NJ. pp 365-386
30. Ryu CM, Murphy CF, Reddy MS, Kloepper JW (2007) A two strain mixture of rhizobacteria elicits induction of systemic resistance against *Pseudomonas syringae* and Cucumber mosaic virus coupled to promotion of plant growth on *Arabidopsis thaliana*. J Microbiol Biotechnol 17:280-286
31. Sadeghi A, Karimi E, Dahazi PA, Javid MG, Dalvand Y, Askari H (2012) Plant growth promoting activity of an auxin and siderophore producing isolate of *Streptomyces* under saline soil condition. World J Microbiol Biotechnol 28:1503-1509
32. SAS Institute Inc. 2013. SAS/STAT® 12.3 User's Guide. Cary, NC
33. Sharma PD, Singh M (2004) Problems and prospects of Organic farming. Bull Indian Soc Soil Sc 22:14-41
34. Singh Y, Singh B, Ladha JK, Khind CS, Khera TS, Bueno CS (2004) Effects of residue decomposition on productivity and soil fertility in rice-wheat rotation. Soil Sci Soc Am J 68:854-864
35. Shaukat K, Affrasayab S, Hasnain S (2006) Growth responses of *Triticum aestivum* to plant growth promoting Rhizobacteria used as a biofertilizer. Res J Microbiol 1:330-338
36. Singh PP, Shin YC, Park CS, Chung YR (1999) Biological control of *Fusarium* wilt of cucumber by chitinolytic bacteria. Phytopathology 89:92-99
37. Spaepen S, Vanderleyden J, Remans R (2007) Indole-3-acetic acid in microbial and microorganism-plant signaling. FEMS Microbiol Rev 31:425-448
38. Tokala RK, Strap JL, Jung CM, Crawford DL, Salove MH, Deobald LA, Bailey JF, Morra MJ (2002) Novel plant-microbe rhizosphere interaction involving *Streptomyces lydicus*WYEC108 and the pea plant (*Pisum sativum*). Appl Environ Microbiol 68:2161-2171

39. Trejo-Estrada SR, Paszczynski A, Crawford DL (1998) Antibiotics and enzymes produced by the biocontrol agent *Streptomyces violaceusniger* YCED-9. J Ind Microbiol Biotechnol 21:81-90

chapter 5

The Use of Legumes as a Biogas Substrate - Potentials for Saving Energy and Reducing Greenhouse Gas Emissions through Symbiotic Nitrogen Fixation

P. Walter Stinner

DBFZ Deutsches Biomasseforschungszentrum, Torgauer Str. 116, Leipzig, 04347, Germany

Justus-Liebig-Universität, Professur für organischen Landbau, Karl-Glöckner-Str. 21C, Gießen, 35394, Germany

ABSTRACT

Background

Energy crops are of considerable importance for biogas production, especially in Germany. The main energy crops for that purpose are corn silage, grass silage, whole crop grain silage and other non-legume crops. The reason for preferring these crops is their high yield, which not only results in high yields of biogas per hectare but also in a high mitigation of greenhouse gases in the course of replacing fossil energy. This article aims to show an additional effect exerted on energy yield and mitigation of greenhouse gases by the use of legume energy crops. The symbiotic nitrogen fixation (SNF) of legumes compensates inorganic N fertilizer in conventional farms, if the digestate is applied as a fertilizer to the non-legume cash crops. The production of chemical N fertilizer is very energy intensive and leads to emissions of greenhouse gases from fossil energy consumption and from nitrous oxide generation. So, the creation of an effective organic fertilizer with nitrogen from biological N_2 fixation is a further energy add-on effect to the reduction of greenhouse gas emissions.

Methods

For this article, data with regard to the SNF of legumes obtained in field experiments at the research station at Gladbacherhof (University of Giessen) from 2002 to 2005 were re-calculated and compared with data concerning energy need and greenhouse gas emissions in the process of producing mineral nitrogen fertilizer. In addition to the possible methane yield of these substrates, the saving in energy and greenhouse gas emissions by substituting mineral fertilizers is shown.

Results

As a result, the possible replacement of primary energy by SNF of clover grass leys is calculated to be approximately less than 6.4 MWh ha^{-1} a^{-1}. This is a yield that is reached in addition to the methane production, i.e. a possible reduction of greenhouse gas emissions through SNF per hectare of clover grass leys of more than 2 t CO_2 equivalents ha^{-1} a^{-1} can be achieved.

Conclusions

Based on these results, it can be recommended to evaluate energy crops in a more holistic way. For legumes, the effect of SNF needs to be included into the energy and greenhouse balance.

BACKGROUND

Using renewable energy sources like biogas can substitute fossil sources and avoid greenhouse gas emissions. Energy crops are of considerable importance for biogas production, especially in Germany [1]. The main energy crops for that purpose are corn silage, grass silage, whole crop grain silage (grain crops chopped in the dough-ripe stage for ensiling) and other non-legume crops. The reason for preferring these crops is their high dry matter yield [2], which results in high yields of biogas per hectare and in high mitigation of greenhouse gases in the process of replacing fossil energy (natural gas, crude oil, lignite and hard coal).

This article aims to show an additional influence exerted on energy yield and mitigation of greenhouse gases through the use of legume energy crops. Via anaerobic digestion, the grown energy crops are transferred into organic fertilizers. In this way, the biological N_2 fixation (symbiotic nitrogen fixation (SNF)) of legumes compensates inorganic N fertilizer in conventional farms, if the digestate is applied as a fertilizer to the non-legume cash crops. The production of chemical N fertilizer is very energy intensive and leads to emissions of greenhouse gases through the utilization of fossil fuels and the generation of nitrous oxide emissions from production processes. In this way, both climate relevant emissions, caused by the use of fossil energy sources in the manufacturing of mineral fertilizer as well as by the generation of nitrous oxide emissions during the production process, could be avoided. Thus, the creation of an effective organic fertilizer with the nitrogen from the biological N_2 fixation is a further energy add-on effect leading to a reduction of greenhouse gas emissions.

METHODS

The data about the biological N_2 fixation of legume crops for biogas substrate derived from field experiments using organic cropping systems. The cropping systems were (i) a typical crop rotation for stockless organic farming systems and (ii) a typical crop rotation for organic dairy farms. The crop rotation of the stockless organic farming systems included: 1. clover grass (CG); 2. potatoes (*Solanum tuberosum L.*); 3. winter wheat (*Triticum aestivum L.*) + catch crop (CC); 4. spring peas (*Pisum sativum L.*) + CC; 5. winter wheat (*T. aestivum*) + CC; and 6. spring wheat (*T. aestivum*) with under-sown CG. The used forage mixture of the CG consisted of *Trifolium pratense L.* at a seed rate of 7.5 kg ha^{-1}, *Medicago sativa L.* at 7.5 kg ha^{-1}, *Lolium perenne L.* at 5 kg ha^{-1},*Phleum pratense L.* at 5 kg ha^{-1} and *Festuca pratensis Huds.* at 5 kg ha^{-1}. The CC mixture consisted of common vetch (*Vicia sativa L.*) at a seed rate of 90 kg ha^{-1} and oil radish (*Raphanus sativus L.*) at 5 kg ha^{-1} (for more details, see [3]).

The crop rotation of the organic dairy farming system included: (1 and 2) 2 years of CG, 3. winter wheat (*T. aestivum*) + CC, 4. corn (*Zea mays L.*) at 80% of the area and potatoes (*S. tuberosum*) at 20% of the area, 5. winter rye (*Secale cereale L.*) + CC, 6. spring peas (*P. sativum*) + CC, 7. spelt (*Triticum aestivum subsp. spelta*) and 8. spring wheat (*T. aestivum*) with under-sown CG. The seed mixtures of the forage ley and the catch crop mixture were the same as described above (for more details, see [4]).

The experiments were carried at the research station of the University of Giessen, Gladbacherhof, from 2002 to 2005. The site is located in the Limburg Basin (latitude 50° 24′ N, longitude 8° 15′ E, altitude: ca. 210 m). The soils are silty loams derived from loess with pH values of 6.6 to 6.9. They are classified as calcic luvisols with a field capacity of 330 to 370 mm m^{-1}. The mean soil C*t*content in the 0- to 30-cm layer was 1.06%, and the N*t* content was 0.125% as described in [5].

The SNF of the CC was assessed by subtracting the soil borne N supply from the total CC N uptake. The soil borne N supply to CC was evaluated according to the data obtained from other trials carried out at the same site using non-leguminous CC (for more details, see [5]). The SNF of the CG was estimated according to the algorithms of Stein-Bachinger et al. [6] by multiplying DM yield (t ha^{-1}), the legume content in the CG (%), and an estimated N derived from the atmosphere (Ndfa) (%). The percentage of Ndfa depends on DM yield, legume content and the above-ground biomass management (harvested/mulched), according to the algorithms provided by the model, see [6]. For more details, see [3].

The cropping of such legume cultures for biogas production is independent of the farming system, the organics or the conventional approach. It can also be carried out in conventional farms and will result in a replacement of non-legume energy crops and inorganic nitrogen fertilizer. The data obtained for the nitrogen yield in these experiments performed in organic farming systems are therefore transferred to conventional systems as an additional N yield, which can replace mineral N fertilizer.

To calculate the savings in energy and greenhouse gases by replacing mineral fertilizer, the occurrence of ammonia losses in the fertilization with digestate, measured in the field experiment, was taken into consideration. According to the investigations of Möller and Stinner [7] in the same research project, the losses amounted to 7% of nitrogen in the case of a separate application of the liquid and solid fermentation products (see column and line 2 of Table 1). After deducting the ammonia losses, the N efficiency of the separated liquid digestate, rich in ammonia is estimated to be 76%, whereas the N efficiency of the separated solid digestate with a high content of organic nitrogen is estimated to be 63% (Table 1). The medium N efficiency of mineral fertilizer is taken as 86% [8]. Thus, on average, 1 kg of digestate N counts for 800 g of mineral fertilizer N.

Table 1: Nitrogen efficiency of digestate and inorganic fertilizer

Fertilizer type	Ammonia losses after spreading (% of total N)	Ammonia content before spreading (% of N*t*)	Effective ammonia content (% of N*t*) after deducting ammonia losses	Overall efficiency (after deducing ammonia losses)	Equivalent to mineral N fertilizer (%)
Separated digestate (liquid phase)	7	71	64	76	88
Separated solid digestate (average of catch crops and clover grass)	7	22	15	63	73
Inorganic nitrogen (mineral fertilizer)	0	-	-	86	1

Stinner

Stinner *Energy, Sustainability and Society* 2015 **5**:4 doi:10.1186/s13705-015-0034-z

These assumptions are based on the averaged content of ammonia and organic nitrogen in the digestate. The ammonia N availability is equivalent to the mineral fertilizers with regard to ammonia volatilization. The long-term availability of the organic bound nitrogen from organic fertilizers amounted to 58% to 59% [8]. For mineral N fertilizer, the author obtained an N efficiency of 86% (see line four, column five in Table 1). This resulted in an equivalence factor of 80% between the mineral fertilizer and the average of the solid and liquid digestate.

The assumptions for digestate can be seen as conservative ones, when new measures of digestate fertilizing (e.g. strip tillage with an underlaying digestate depot) will be optimized and proved (for efficient cultivation options, see Hauggaard-Nielsen et al. [9]).

The values according to Nemecek and Kägi [10],[11] and Nemecek and Schnetzer [12] have been used to calculate the savings in energy and greenhouse gases when replacing the mineral fertilizer nitrogen (see also [13]). In the database for life-cycle analyses of agricultural production systems, an energy consumption for common nitrogen mineral fertilizer (ammonium nitrate) of 58.2 MJ × kg N^{-1} and emissions of 6.032 kg CO_2 eq. kg N^{-1} was obtained.

RESULTS AND DISCUSSION

Table 1 in the 'Methods' section takes not only into account the ammonia losses after spreading of digestate but also the different nitrogen efficiencies of organic and inorganic fertilizers. With regard to [13] and [7], the solid–liquid separation of digestate is a good practice measure to reduce the ammonia losses after spreading of the digestate. Due to this method, the nitrogen losses in the whole process chain as well as the N efficiency on the field are of great importance for the results. Important causes of losses are ammonia volatilizations in the whole process chain. In both parts, in the biogas plant and in field application, these losses are strongly influenced by the technique and management used [14]-[20]. The field N efficiency also depends on a couple of further factors in addition to the kind of fertilizer, especially in the case of organic fertilizers. These factors are application technique, time and rate of application, fertilized crop and nutrient management by crop rotation and cropping methods [21]-[25].

The field experiment shows an average SNF of 122 kg N ha^{-1} (Table 2), see also [3], for arable land with the given crop rotation. One hectare of clover grass leys as a biogas substrate can produce up to 500 kg of nitrogen through SNF per main year of cultivation ([13], compared with[26]). The average SNF of legume catch crop mixtures amounted to 61 kg N per hectare of cover cropping. These amounts are effects achieved in addition to the potential of methane production from the harvested crops. The nitrogen in the investigated ecological system was produced without restrictions in food production. It seems to be possible to transfer the results to conventional systems. In the case of the clover grass leys, the additional nitrogen harvest as well as further effects (e.g. humus balance, preceding crop effects) needs to be balanced with lower direct biomass yields compared to other substrate crops (e.g. corn silage, whole crop silage). For the humus balance, the cropping of legume forage lays (without including the digestate) generated a surplus of 600 to 800 humus equivalent units per hectare of main cropping depending on the yield level [27]. This surplus can cover the humus needs of other crops in the crop rotation. In contrast to that, non-legume crops for energy production or other purposes (also without taking the digestate into consideration) showed a need of 560 (corn silage), 760 (sugar beet) or 280 humus equivalent units (cereals, threshed, minimum values, respectively, see also [27]). This considerable advantage of legume forage leys should be taken into account, when comparing different energy crops.

Table 2: Energy consumption [kWh ha^{-1}] and greenhouse gases emissions [kg CO2eq. ha^{-1}] of N fertilizers

Fertilizer type	Non-renewable energy input per kg of produced nitrogen [kWh/ kg N]	Greenhouse gases emissions during production [t CO2eq. ha−1]	N2O emissions after application [kg CO2eq. kg−1of applied fertilizer]	N2O emissions after application [kg CO2eq. kg−1of inorganic N fertilizer equivalent]	Sum of greenhouse gas emissions [kg CO2eq. kg−1of inorganic N fertilizer equivalent]
Digestate, N derived from SNF[a]	-	-	2.98[b]	3.73	3.73
Inorganic nitrogen	16.17	6.032	2.98[b]	2.98	9.01

[a]Fuel and energy for biogas plant operation belongs to methane production. [b]According to the IPCC scheme [10],[12], the N_2O emissions after application are similar but due to the lower efficiency, a higher amount is needed. So, the relevant value is the emission per kg of replaced mineral fertilizer (next column).

Stinner

Stinner *Energy, Sustainability and Society* 2015 5:4, doi:10.1186/ s13705-015-0034-z

Table 2 shows the savings in non-renewable energy (fossil + nuclear) and greenhouse gas emissions per kg N through SNF when replacing mineral fertilizer by digestate. The value for the transfer of N to N_2O is the same for inorganic and organic fertilizers [12]. However, the N efficiency of the digestate is only calculated to reach 80% of the mineral fertilizers efficiency (average of liquid and solid fraction, see Table 1), i.e. a higher amount of digestate N for the same fertilizing effect. Due to the fact that N_2O generation is calculated on the basis of the applied N, a higher transfer of digestate nitrogen into nitrous oxide after fertilization is calculated to be 2.98: 0.8 = 3.73, where 2.98 is the N_2O emission after the application in kg of CO_2 equivalents for both kinds of fertilizer. The divisor 0.8 derives from the equivalence factor between digestate N and mineral N, as shown in the Methods. The resulting value denotes the digestate emissions of 3.73 kg CO_2 equivalents per kg of mineral N, which is replaced by (1.25 kg of) digestate N. In this way, the results of digestate emissions are comparable to the values of mineral fertilizer.

The net savings of greenhouse gases through SNF [kg CO_2 eq. kg^{-1} of inorganic N fertilizer replaced] are 9.01 − 3.73 = 5.28, where 9.01 is the sum of greenhouse gas emissions (in kg of CO_2equivalents) per kg of N caused by mineral fertilizer and 3.73 is the greenhouse gas emissions (in kg of CO_2 equivalents) for the replacement of 1 kg inorganic N by digestate (Table 2).

The energy for cropping and processing the clover grass is calculated as part of the methane production as well as the emissions from cropping and operating the biogas plant. This makes the values comparable with the use of other energy crops without the benefits from SNF, such like corn silage.

Table 3 shows the values of energy savings and greenhouse gases calculated for the given field data. In the case of the catch crops, the results of the field experiments did not show any differences in the dry matter yield regarding the N nutrition [5]. Depending on the N availability, only the composition of the mixture was adapted. Higher N availability leads to higher contents of the non-legume partner in the mixture and vice versa. This effect was often found by other authors as cited in [5] and [13]. This may be different, if other catch crops can be integrated (e.g. annual ryegrass). The nitrogen yields by SNF were strongly dependent on the nitrogen availability: as higher the nitrogen availability, the lower the SNF. This means, a strategy to optimize nitrogen harvest should be to only recognize the use of fertilizers of a wide C/N ratio in the fertilization of legume mixtures for biogas substrate. This is evident when comparing the SNF after cereals and peas (Table 3) and furthermore when studying the results of catch crops amended with digestate (data not shown).

Table 3: Savings of energy and greenhouse gases by replacing mineral nitrogen

Element of cultivation	SNF	**Replacement of mineral fertilizer**	**Savings of energy [kWh ha−1]**	**Savings of greenhouse gases [kg CO2eq. ha−1]**
Clover grass leys as catch crop (CC, autumn growth after harvest of spring wheat)	65	52	841	275
Clover grass leys (year of main crop)	498	398	6,436	2,101
CC after WW3	77	62	1,003	327
CC after peas	36	29	469	153
CC after WW5	66	53	857	280
$\overline{X}$ CC	61	49	792	259
arable land (AL, 6 ha)	730	584	9,443	3,084
$\overline{X}$ AL	122	98	1,578	517

Acronyms: WW3 = winter wheat in third place of crop rotation, WW5 = winter wheat in fifth place of crop rotation.

Stinner

Stinner *Energy, Sustainability and Society* 2015 **5**:4, doi:10.1186/s13705-015-0034-z

Depending on the different previous crops and the remaining nitrogen, the catch crops have fixed 36 to 77 kg N ha^{-1} of nitrogen by SNF (x^{-} = 61 kg N ha^{-1}, Table 1). For most of the locations relating to catch crops, as determined by water supply and the harvest period of the previous crops, a transfer to conventional conditions is possible. To ensure a high SNF, the use of nitrogen fertilizers to legume mixtures must be avoided because high nitrogen availability leads to higher contents of non-legume partners in the mixtures and a low SNF from legumes [5],[28]-[32].

This means that a sufficient capacity of fermentation residue storage is necessary. For high SNF, forage mixtures or catch crops with the option to be mixtures with legumes should not be regarded as an element for clearing the digestate storages anymore. Alternative adjustments in the selection and management of non-legume market crops are required. With the usual technique, the fertilization of winter grains with liquid digestate (of a narrow C/N-ratio) should be carried out under the beneficial soil conditions at the beginning of the vegetation period in early spring. Concentrating injection techniques, optionally combined with measures against nitrification, have proved to be suitable for the fertilization of such crops with liquid organic fertilizers. For instance, in the cultivation of corn, the fertilization with digestate would be possible until July, provided an appropriate equipment is applied. The cash crop for a moderate fertilization with digestate (or other organic fertilizers of a small C/N ratio) in the second half of the year is rapeseed (*Brassica napus L.*) due to its high nitrogen uptake before wintertime [33]. This crop can be amended with digestate ideally before seeding in August. The amount of digestate for a reasonable fertilization strongly depends on the N surpluses and the straw management of the pre-crop.

In addition to the possible biogas production [34], the use of clover grass leys as a biogas substrate can reduce the primary energy consumption to more than 6.4 MWh ha^{-1} and the greenhouse gas emissions to more than 2 t CO_2 eq. ha^{-1} (Table 2) in a conventional system when replacing mineral nitrogen fertilizer. Legume intercrop mixtures were able to save approximately 0.8 MWh ha^{-1} a^{-1} and more than 200 kg CO_2 equivalents ha^{-1} a^{-1}.

At the moment, in conventional systems, the use of such legume mixtures for biogas production is almost of no significance. Corn silage and further non-legumes are the most important crops producing biogas in Germany [1] due to their high methane yields. Indeed, using grass leys as substrate crops is an interesting alternative under certain local conditions [35].

The results show additional effects exerted by legume energy crops on energy yields and greenhouse gas savings, which have not been discussed by other authors until now. The SNF, which is set up above, has led to an improvement of this crop.

Further advantages of applying legume forage leys as energy crops are the following: expansion of tight crop rotations, abatement of the risks for the corn cultivation through the corn rootworm[36]-[38], extended soil regeneration period, protection against erosion, lower production of nitrous oxide by the soil as compared to the use of non-legumes due to the absence of nitrogen fertilization [7],[10], and lower danger of nitrate leaching through soil coverage all year. The high competitive strength of clover grass leys is a measure of good weed inhibition. Considering nitrogen efficiency and the mentioned agronomic effects [39],[40], the economic use of clover grass leys and catch crop legumes when applying a biogas technique may be of advantage also under conventional circumstances.

Under this background, the efficiencies of different energy crops should be re-discussed and compared to the literature findings [2],[41],[42].

CONCLUSIONS

The results show a strong underestimation of the yields of legumes as energy crops. Currently, energy crops for biogas digestion are mainly annual non-legume crops such as corn silage, whole crop grain silage (grain crops chopped in the dough-ripe stage for ensiling), etc. These crops are mainly regarded as the most effective because of their high methane production per hectare and the great replacing of fossil energy carriers. The high amount of energy saving (e.g. more than 20% of energy in addition to the methane yield of clover grass leys [13],[43]) and the greenhouse gas mitigation by replacement of mineral nitrogen by symbiotic fixed nitrogen of legume crops for biogas production are remarkable, when the digestate is used to fertilize the non-legume crops. These results should be integrated in holistic evaluations of energy crops. This measure may change the relative advantages of different crops as biogas substrates.

ACKNOWLEDGEMENTS

I would like to thank the Deutsche Bundesstiftung Umwelt (DBU) for the financial support of the field experiments, Dr. Kurt Möller for creating the research concept as well as for a lot of helpful discussions and joint publications. Furthermore, I am grateful to my PhD colleague Arno Deuker for all the work done together, the colleagues from the chair of organic agriculture including the research station as well as the colleagues of the chair of microbiology of the University of Giessen for all the support on field, in lab and in discussions of the results in the area of soil borne trace gas emissions. I would like to thank Prof. Dr. Günter Leithold for opening his research capacities for the combination of biogas and organic farming. Thanks are due to my colleagues from the DBFZ for supporting the *ecoinvent* database and for reading the script, in particular, to Elisa Thieme, Jaqueline Daniel-Gromke, Velina Denysenko, Eva Nebel, Katja Oemichen and Stefan Majer. And last but not least, I would like to thank Dagmar Fiedler (UFZ) for proofreading the manuscript.

REFERENCES

1. DBFZ (2014) Stromerzeugung aus Biomasse; (Vorhaben IIa Biomasse), Zwischenbericht Juni 2014; Projektleitung: Mattes Scheftelowitz; Projektbearbeitung: Jaqueline Daniel-Gromke, Nadja Rensberg, Velina Denysenko, Konrad Hillebrand, Karin Naumann, Mattes Scheftelowitz, David Ziegler, Janet Witt (DBFZ); Michael Beil, Wiebke Beyrich (beide IWES); available: https://www.dbfz.de/web/presse/pressemitteilungen-2014/dbfz-veroeffentlicht-monitoring-zwischenbericht-zur-stromerzeugung-aus-biomasse.html
2. Vetter A, Eckner J, Strauß C, Nehring A (2014) Entwicklung und Optimierung von standortangepassten Anbausystemen für Energiepflanzen im Fruchtfolgeregime; Abschlussbericht zum Teilprojekt 1; (EVA II); available: http://www.eva-verbund.de/untersuchungsberichte/teilprojekte/teilprojekt-1-pflanzenbau.html
3. Stinner W, Möller K, Leithold G (2008) Effects of biogas digestion of clover/grass-leys, cover crops and crop residues on nitrogen

cycle and crop yield in organic stockless farming systems. Eur J Agron 29:125-134 Publisher Full Text

4. Möller K, Stinner W, Deuker A, Leithold G (2008) Effects of different manuring systems with and without biogas digestion on nitrogen cycle and crop yield in mixed organic dairy farming systems. Nutr Cycl Agroecosyst 82:209-232
5. Möller K, Stinner W, Leithold G (2008) Growth, composition, biological N_2 fixation and nutrient uptake of a leguminous cover crop mixture and the effect of their removal on field nitrogen balances and nitrate leaching risk. Nutr Cycl Agroecosys 82:233-249
6. Stein-Bachinger K, Bachinger J, Schmitt L (2004) Naehrstoff-Management im oekologischen Landbau. In: KTBL, Darmstadt (Ed.), Ein Handbuch fuer Beratung und Praxis. Landwirtschaftsverlag Muenster, Muenster, pp 109–114
7. Möller K, Stinner W (2009) Effects of different manuring systems with and without biogas digestion on soil mineral nitrogen content and on gaseous nitrogen losses (ammonia, nitrous oxides). Europ J Agron 30:1-16
8. Körschens M (1987) N-Ausnutzung in Abhängigkeit von mineralischer und organischer N-Düngung im Verlaufe von vier Jahrzehnten im statischen Düngungsversuch Lauchstädt. Archiv. Acker Pflanzenbau Bodenkunde 31:161-168
9. Hauggaard-Nielsen H, Johansen A, Carter MS, Ambus P, Jensen ES (2013) Annual maize and perennial grass-clover strip cropping for increased resource use efficiency and productivity using organic farming practice as a model. Eur J Agron 47:55-64
10. Nemecek T, Kägi T (eds) (2007) Life Cycle Inventories of Agricultural Production Systems; Ecoinvent report No. 15, Agroscope Reckenholz – Tänikon Research Station ART, Zürich and Dübendorf, http://www.upe.poli.br/~cardim/PEC/Ecoinvent%20LCA/ecoinventReports/15_Agriculture.pdf. Accessed 17 June 2014
11. Database ecoinvent 3 (2014), https://ecoquery.ecoinvent.org/Account/LogOn?Return Url=%2 f. Accessed 19 June 2014
12. Nemecek T, Schnetzer J (2011) Methods of assessment of direct field emissions for LCIs of agricultural production systems.

Agroscope Reckenholz – Tänikon Research Station ART, Zürich and Dübendorf.

13. Stinner W (2011) Auswirkungen der Biogaserzeugung in einem ökologischen Marktfruchtbetrieb auf Ertragsbildung und Umweltparameter; Giessener Schriften zum ökologischen Landbau; Band 4. Verlag Dr. Köster, Berlin.
14. Daniel-Gromke J, Liebetrau J, Denysenko V, Krebs C (2014) Digestion of bio-waste - GHG emissions and mitigation potential (accepted for publication in Energy, Sustainability and Society)
15. Huijsmans JFM, Hol JMG, Hendriks MMWB (2001) Effect of application technique, manure characteristics, weather and field conditions on ammonia volatilization from manure applied to grassland. NJAS - Wageningen J Life Sci 49(4):323-342
16. Huijsmans JFM, Hol JMG, Vermeulen GD (2003) Effect of application method, manure characteristics, weather and field conditions on ammonia volatilization from manure applied to arable land. Atmos Environ 37(26):3669-3680
17. Kirchmann H, Witter E (1989) Ammonia volatilization during aerobic and anaerobic manure decomposition. Plant Soil 115:35-41
18. Bernal MP, Kirchmann H (1992) Carbon and nitrogen mineralization and ammonia volatilization from fresh, aerobically and anaerobically treated pig manure during incubation with soil. Biol Fertil Soils 13(3):135-141
19. Bussink DW, Oenema O (1998) Ammonia volatilization from dairy farming systems in temperate areas: a review. Nutr Cycl Agroecosyst 51:19-33
20. Huijsmans JFM, de Mol RM (1999) A model for ammonia volatilization after surface application and subsequent incorporation of manure on arable land. J Agric Eng Res 74(1):73-82
21. Fageria NK, Baligar VC (2005) Enhancing nitrogen use efficiency in crop plants. Adv Agron 88:97-185
22. Pierce FJ, Rice CW (1988) Crop rotation and its impact on efficiency of water and nitrogen use. In: Ellis BG, Hargrove WL (eds) Cropping strategies for efficient use of water and nitrogen, vol 51. ASA Special Publications, American Society of Agronomy, pp 21–42

23. Dawson JC, Huggins DR (2008) Jones SS (2008) Characterizing nitrogen use efficiency in natural and agricultural ecosystems to improve the performance of cereal crops in low-input and organic agricultural systems. Field Crop Res 107:89-101
24. Edmeades DC (2003) the long-term effects of manures and fertilisers on soil productivity and quality: a review. Nutr Cycl Agroecosyst 66:165-180
25. Smith KA, Chambers BJ (1993) Utilizing the nitrogen content of organic manures - on-farms-problems and practical solutions. Soil Use Manag 9:3
26. Richards JE, Soper RJ (1979) Effect of N fertilizer on yield, protein content, and symbiotic N fixation in faba beans. Agron J 71:807-811
27. VDLUFA, Hrsg, (2014) Standpunkt; Humusbilanzierung - Eine Methode zur Analyse und Bewertung der Humusversorgung von Ackerland; Verband Deutscher Landwirtschaftlicher Untersuchungs- und Forschungsanstalten; available: http://www.vdlufa.de/index.php?option=com_content&view=article&id=21&Itemid=37 (30.12.2014)
28. Brozyna MA, Petersen SO, Chirinda N, Olesen JE (2013) Effects of grass-clover management and cover crops on nitrogen cycling and nitrous oxide emissions in a stockless organic crop rotation. Agr Ecosyst Environ 181:115-126
29. Jensen ES (1987) Seasonal patterns of growth and nitrogen fixation in field-grown pea. Plant Soil 101:29-37
30. Hansen JP, Vinther FP (2001) Spatial variability of symbiotic N_2 fixation in grass-white clover pastures estimated by the 15 N isotope dilution method and the natural 15 N abundance method. Plant Soil 230:257-266
31. Hardarson G, Atkins C (2003) Optimizing biological N_2 fixation by legumes in farming systems. Plant Soil 252:41-54
32. Høgh-Jensen H, Schjørring JK (1997) Interactions between white clover and ryegrass under contrasting nitrogen availability: N_2 fixation, N fertilizer recovery, N transfer and water use efficiency. Plant Soil 197(2):187-199
33. Rossato L, Lainé P, Ourry A (2001) Nitrogen storage and remobilization in *Brassica napus* L. during the growth cycle:

nitrogen fluxes within the plant and changes in soluble protein patterns. J Exp Bot 52(361):1655-1663

34. Deuker A, Stinner W, Leithold G (2008) Biogas energy potentials from agricultural by-products: examples from organic farming in Germany compared with energy corn. In: Abstracts of the 16th European Biomass Conference and Exhibition - From Research to Industry and Markets, Valencia, Spain, 2–6 June 2008; ETA-Florence Renewable Energies, Florence (ed.)
35. FNR (2011) Standortangepasste Anbausysteme für Energiepflanzen. Gülzow, Germany, http://mediathek.fnr.de/media/downloadable/files/samples/f/g/fg_broschur_energiepfl_eva_2011_v2.pdf. Accessed 17 June 2014
36. Hummel HE, Deuker A, Leithold G (2009) the leaf beetle Diabrotica virgifera virgifera LeConte: a merciless entomological challenge for agriculture. In: Tasin M, Witzgal P (eds) IOBC-WPRS Bulletin, vol 41, p 103–110
37. Deuker A, Stinner W, Rensberg N, Wagner L, Hummel HE (2012) Regional risks for the biogas production in Germany by the maize pest *Diabrotica v. virgifera*. J Agr Sci Tech 2:749-767
38. Deuker A (2013) Energieerzeugung in landwirtschaftlichen Biogasanlagen: Potenziale und Grenzen. Verlag Dr. Köster, Berlin.
39. Kramberger B, Gselman A, Kristl J, Lešnik M, Šuštar V, Muršec M, Podvršnik M (2014) Winter cover crop: the effects of grass-clover mixture proportion and biomass management on maize and the apparent residual N in the soil. Eur J Agron 55:63
40. Frøseth RG, Bakken AK, Bleken MA, Riley H, Pommeresche R, Thorup-Kristensen K, Hansen S (2014) Effects of green manure herbage management and its digestate from biogas production on barley yield, N recovery, soil structure and earthworm populations. Eur J Agron 52:90-102
41. Boehmel C, Lewandowski I, Claupein W (2008) comparing annual and perennial energy cropping systems with different management intensities. Agric Syst 96(1):224-236
42. Launder K (2002) Energy crops and their potential development in Michigan; Michigan Biomass Energy Program, August, 2002; Michigan Department of Consumer and Industry Services, Energy Office (ed)

43. Keymer U (2014): Biogasausbeuten verschiedener Substrate; available under http://www.lfl.bayern.de/iba/energie/049711/ (01.12.2014)

Chapter 6

Priority areas for watershed service conservation in the Guapi-Macacu region of Rio de Janeiro, Atlantic Forest, Brazil

Vanesa Rodríguez Osuna[1,2], Jan Börner[1], Udo Nehren[2], Rachel Bardy Prado[3], Hartmut Gaese[1], and Jürgen Heinrich[4]

[1]Center for Development Research (ZEF), Department of Economic and Technological Change, University of Bonn, Walter-Flex-Str. 3, Bonn, 53117, Germany

[2]Institute for Technology and Resources Management in the Tropics and Subtropics (ITT), Cologne University of Applied Sciences, Betzdorfer Str. 2, Cologne, 50679, Germany

[3]Empresa Brasileira de Pesquisa Agropecuária (EMBRAPA Soils), Brazilian Enterprise for Agricultural Research, Geotechnologies and Environmental Monitoring, Rua Jardim Botânico, Rio de Janeiro, CEP 22460-000, Brazil

[4]Institute for Geography, Department of Physical Geography, University of Leipzig, Johannisallee 19a, Leipzig, 04103, Germany

ABSTRACT

Introduction

Land use intensification and urbanisation processes are degrading hydrological ecosystem services in the Guapi-Macacu watershed of Rio de Janeiro. A proposal to pay farmers to restore natural watershed services might be an alternative to securing the water supply in the long-term for the around 2.5 million urban water users in the study region. This study quantifies the costs of changing current land use patterns to enhance watershed services and compares these costs to the avoided costs associated with water treatment for public supply.

Methods

We use farm-household data to estimate the opportunity costs of abandoning current land uses for the recovery of natural vegetation; a process that is very likely to improve water quality in terms of turbidity due to reduced inputs from erosion. Opportunity cost estimates are extrapolated to the watershed scale based on remote sensing land use classifications and vulnerability analysis to identify priority zones for watershed management interventions. To assess the potential demand for watershed services, we analyse water quality and treatment cost data from the main local water treatment plant.

Results

Changing agricultural land uses for watershed services provision generally comes at high opportunity costs in our study area near to the metropolis of Rio de Janeiro. Alternative low cost watershed conservation options do exist in the livestock production sector. These options have the potential to directly reduce the amount of sediments and nutrients reaching the water bodies, and in turn decrease the costs of treatment needed for drinking water. Land cover changes at the scale needed to improve water quality will, nonetheless, likely exceed the cost of additional investments in water treatment.

Conclusions

The state water utility company's willingness to pay for watershed services alone will not be enough to induce provision of additional watershed services. We conclude that monetary incentives conditioned on specific adjustments to existing production systems could still have a complementary role to play in improving watershed services. However, we note that our willingness to pay analysis focusses on only one of the potentially wide range of ecosystem services provided by natural vegetation in the Guapi-Macacu watershed. Factoring these ecosystem services into the willingness to pay equation is likely to change our assessment in favour of additional conservation action, be it through PES or other policy instruments.

INTRODUCTION

One of the biggest challenges of our time is to revert the ongoing degradation of ecosystems while meeting the increasing demand for food and biomass (Thomas and Callan [2010]; MEA [2005]). Population and economic growth are increasing the demand for water resources and, at the same time, amplifying the pressure on ecosystems that deliver watershed services (Porras et al.[2008]).

The degradation of ecosystem services represents a loss of widely undervalued natural capital assets (TEEB [2009]; Montes [2007]); MEA [2005]. While there is general agreement that land use choices influence watershed service provision, the magnitude and nature of the effects are highly context dependent and poorly understood in tropical forest environments (Porras et al. [2008]; Veiga [2008]; Calder [2005]; van Noordwijk [2005]).

Land and water linkages are challenging to manage as watershed services have a common good characteristic (Porras et al. [2008]) and are generally unaccounted for in monetary terms. As a result, they are often degraded and this is not reflected in traditional economic measures, such as GDP (TEEB [2009]; Costanza et al. [1997]; FAO [2007]). Land use patterns and land management practices by upstream landholders in a watershed determine, to a great extent, quality and quantity of water bodies (Porras et al. [2008]). For example, unsustainable land use and agricultural practices can result in negative hydrological side-effects, or externalities, such as sedimentation (Veiga [2008]; MEA [2005]). As a result, the design of policy measures that encourage farmers to adopt watershed conservation measures and, at the same time, maintain or even increase productivity has become a major research issue.

Several policy options are available to enhance incentives for the supply of ecosystem services. Command-and-control regulations, such as bans on forest clearing, have long been the preferred policy option to control environmentally harmful land use changes (Boerner and Vosti [2012]; Porras et al. [2008]; FAO [2007]). More recently, mechanisms based on economic incentives or markets for ecosystem services are being proposed as cost-effective complementary tools to promote ecosystem service conservation (Porras et al. [2008]; Engel et al. [2008]; FAO [2007]). These incentive-based mechanisms include payments for ecosystem services (PES) and payments for watershed services (PWS), water quality trading markets and reciprocal or in-kind agreements (Bennett et al. [2013]).

Latin America registered a transaction value of USD 528.9 million in such investments between 2001 and 2011 corresponding to 3.4 million hectares (Bennett et al. [2013]). PWS programmes are considered to be the "fastest growing" and most mature among all PES schemes in Latin America (Bennett et al. [2013]; Balvanera et al. [2012]; Stanton et al. [2010]). PWS in Latin America typically involve compensating upstream rural producers for protecting and/or restoring natural forest ecosystems and highland natural pastures (páramo) (Pria et al. [2013]; Veiga and Galvadao[2011]; Grieg-Gran et al. [2005]). Such ecosystems are mostly located in strategic water production areas, such as headwaters, riparian forest or water intake points for water supply (Pria et al. [2013]; Veiga and Galvadao [2011]). There is, however, little evidence on the effectiveness of such schemes in meeting conservation and development goals in the peer-reviewed literature (see for example Arriagada et al. [2012]; Pattanayak et al. [2010]).

In Brazil, PWS are expanding and there are already 848 ecosystem service providers in the Atlantic Forest region alone, mainly organised or supported within the National Water Agency's Water Producer Programme (Veiga and Galvadao [2011]; Santos et al. [2010]). As of

2012, 41 PWS projects had been implemented or were in preparation, covering an area of around 40,000 hectares. A project in the state of Rio de Janeiro is currently in development, in the Três Picos State Park, Municipality of Cachoeiras de Macacu (Guedes and Seehusen [2011]). This State Park is located in the mountainous area of the Guapi-Macacu watershed, which contributes to the supply of drinking water for almost 2.5 million inhabitants of five municipalities, including the city of Niteroi (Pedreira et al. [2009]) and this watershed is the focus of the present study. The main drivers of degradation of water resources in this watershed are urbanisation processes, intensive agriculture and conversion of riparian vegetation.

Managing watershed services requires solid knowledge about (1) the costs of providing an additional unit of water quality or quantity (supply side) and (2) water users or intermediaries' willingness to pay for an additional unit of water quality or quantity (demand side). The economic aspects of watershed services supply and demand are particularly poorly studied and many PWS initiatives begin operating without quantitative knowledge of such parameters (Martin-Ortega et al.[2012]). To address this knowledge gap, we focus on two research questions within our study area of the Guapi-Macacu watershed in the state of Rio de Janeiro:

1. What are the costs associated with shifting land uses to foster improved and enhanced watershed services (services supply)?
2. What are the treatment costs for the water supply company to improve water quality parameters such as turbidity?

Our focus is on understanding the supply and environmental conditions that determine watershed service provision. Particularly, we concentrate on the watershed service related to water quality maintenance in terms of turbidity reduction for supplying drinking water. However, we also analyse factors that influence water treatment costs (related to services demand) based on land use changes. The remainder of this paper is structured as follows, section "Assessing watershed services supply and demand" presents our approach, addresses relevant literature and concepts related to watershed service supply and demand, section "Methods" describes the study area and methods used for this research, section "Results" presents our findings, section "Discussion" discusses these findings and contextualises them with other studies and reviews the main caveats. Finally, section "Conclusions" presents our main conclusions and the policy implications for watershed service management.

ASSESSING WATERSHED SERVICES SUPPLY AND DEMAND

The Supply of Watershed Services

Watershed service provision is to a large extent determined by land use and land cover. For example, unsustainable land use is frequently linked to a high surface runoff and an elevated concentration of suspended and soluble loads in water bodies (Batchelor et al. [1998]). Changes in land cover, such as forest to agriculture conversion, tend to increase superficial runoff and sediment flux in rivers (FAO [2007]; MEA [2005]).

Healthy forest and wetland ecosystems are considered very effective at regulating water flow and improving water quality (Russi et al. [2013]; TEEB [2010a] [b]). Maintaining water quality includes the control of sediment, nutrient (in particular phosphorous and nitrogen) and chemical loads, as well as salinity (TEEB [2010b]). In addition, forest ecosystems can remove pathogenic microbes, sequester and convert inorganic ions and transform persistent organic pollutants (TEEB [2010b]).

Watersheds with an important extension of forest tend to offer better water quality than those subject to other land uses, such as agriculture, pasture, industry and urban infrastructure, since the latter are associated with higher discharges of diverse types of pollutants into soils and water. In this way, the presence of forest could substantially reduce the cost of water treatment for drinking water in most cases, thereby reducing the related costs for water supply (Medeiros et al.[2011]; Reis [2004]).

Cities such as Rio de Janeiro, Johannesburg, Tokyo, Melbourne, New York and Jakarta all depend on protected areas with forests to provide drinking water for their residents (Dudley and Stolton[2003]). Moreover, a third of the hundred largest cities worldwide take a significant proportion of their drinking water from protected forested areas (Dudley and Stolton [2003]).

Worldwide several examples show that well-managed forests, wetlands and protected areas are very likely to supply clean water at significantly lower costs than man-made replacements, such as water treatment plants (Bennett et al. [2013]; Hanson et al. [2011]; TEEB [2009]; Postel et al.[2005]). Some examples are included in Table 1. For example, in the United States of America, 27 water suppliers showed that in watersheds with at least 60% forest cover, treatment costs were 50 percent less than those with 30% forest cover (Postel et al. [2005]). Accordingly, several North American cities have decided to invest in watershed management to avoid the expense of water treatment plants.

Table 1: Examples of the economic values of watershed services

Action	Examples	Source
Payments to maintain water purification services in the Catskills watershed, New York avoid the costs of launching a filtration plant for water treatment	Payments for maintaining watershed purification represent 1 to 1.5 billion USD in comparison to the much higher estimated cost of a filtration plant (6–8 billion USD plus 300–500 million USD yearly for operating costs)	Hanson et al. [2011]; WRI[2011]; Pagiola et al.[2004]
The cost of removing nitrate contamination from water in Rochester, Minnesota	2.8-4.8 million USD per year	Hawkins [2003]
WTP for water quality improvement in the levels of iron, sulphate, hardness, and copper in South-western, Minnesota	2.4, 2.0, 6.6 and 2.6 million USD yearly respectively (via contingent valuation method)	Hawkins [2003]
Yearly revenues in public utility resulting from natural water supply in Mud Lake, Minnesota/ South Dakota	94 USD per acre	Hawkins [2003]

Annual WTP from residents living close to the Minnesota River for reducing phosphorous levels	14.1 USD via taxes or 19.6 USD via water bills for a 40% reduction in a nearby river (contingent valuation method)	Hawkins [2003]
Increased reservoir water quality and surface area in St. Louis, Missouri	Yearly net benefit of 25 million USD (travel cost method)	Hawkins [2003]
Value of water supply in Milesburg, Pennsylvania	Between 14 and 36 USD per household (avoided cost)	Hawkins [2003]
In Venezuela, a national protected area system avoids sedimentation	Without the provision of this service by the national protected area, unavoided sedimentation could reduce farmer's income by approximately 3.5 million USD yearly	Pabon-Zamora et al.[2008]
The current provision of ecosystem services related to existing stream vegetation along the Llobregat River, Spain	79,000 EUR per year savings in water treatment costs for the residents of Barcelona	Honey-Roses et al. [2013]
In New Zealand, the value of water provision from the Otago region coming from the Te Papanui Conservation Park	Now it is "free" but would cost 136 million NZD[a] to bring it in from somewhere else (total benefit of the service)	TEEB [2009]

[a]1 USD is equivalent to 1.30 NZD at current exchange rate (http://www.oanda.com/currency/converter/ webcite).

Rodríguez Osuna et al.

Rodríguez Osuna et al. Ecological Processes 2014 3:16, doi:10.1186/s13717-014-0016-7

In the literature regarding water quality, off-site effects of soil erosion are frequently referred to as sedimentation (Veiga [2008]; Holmes [1988]). A summary of economic activities most affected by loss of watershed services is presented in Table 2.

Table 2: Examples of economic activities most affected by the loss of watershed services

Economic activities and damage caused by the loss of watershed services	Examples of incurred costs/ damages due to the loss of watershed services	Source
Activities that depend on reservoirs	Capacity loss for energy generation	Veiga [2008]; Reis [2004]
	Capacity loss of irrigated production	
	Loss of benefits related to flood control	
	Capacity loss of navigation channels	
	Increase of suspended material/ siltation of water bodies resulting in a reduction of their storage capacity	
Drainage and maintenance operations	Costs related to (a) irrigation and drainage of channels, (b) hydroelectric plant reservoirs, (c) ports and (d) eutrophication (increase concentration of N and P in water bodies)	Veiga [2008]; Reis [2004]
Increased water treatment costs related to augmented turbidity	One example are watersheds in São Paulo with lower costs related to consumption of chemical products for the water treating process (less than 20 BRL[a] per 1,000 m^3 of treated water) were those with a forest cover higher than 15%. The two water treatment units with less percentage of forest cover (Piracicaba and Atibaia rivers with less than 10% forest cover) show considerably higher specific costs of chemical products	Reis [2004]; Dearmont et al. [1998]; Holmes [1988], Moore and McCarl [1987]

Monetary damages related to sedimentation	Additional annual costs from 35 to 661 million USD for a US water treatment plant due to soil erosion. These values relate strongly to agricultural production causing sedimentation	López [1997]; Holmes[1988]
	Additional annual operational cost of 3.2 BRL per hectare and 0.10 BRL per ton of eroded soil for the water treatment of the river Corumbataí in São Paulo. These estimates were made assuming that the only externality caused by soil erosion were increased water treatment costs but not taking into account siltation of river banks, flooding of areas close to river banks and loss of navigation capacity	

[a]1 USD is equivalent to 2.38 BRL at current exchange rate (http://www.oanda.com/currency/converter/ webcite). This rate will be used onwards in this document when comparison among monetary units is required.

Rodríguez Osuna *et al.*

Rodríguez Osuna *et al. Ecological Processes* 2014 **3**:16, doi:10.1186/s13717-014-0016-7

Different authors have classified ecosystem services in distinct ways (Haines-Young and Potschin[2013] Daily and Matson [2008]; FAO [2007]; Farber et al. [2006]; MEA [2005]; Postel et al.[2005]; Hawkins [2003]; De Groot et al. [2002]; Costanza et al. [1997]). In the current study, we followed the TEEB ([2010b]) classification of watershed services, whereby maintenance of water quality for human consumption is considered a provisioning service. We use this terminology throughout this article.

It is often argued that a major challenge of market mechanisms relates to the "packaging" of ecosystem services into commodities that are tradable or subject to a contract (Porras et al.[2008]). However, most PWS schemes in developing countries are guided by a "land-based" approach, whereby suppliers are paid to improve their land management practices, which are in turn considered highly likely to result in watershed service provision, rather than being paid for the actual service delivery (Porras et al. [2008]). Regardless of the approach chosen, watershed services supply is inevitably linked to farmers' land use decisions (FAO [2007]). Consequently, watershed services supply analyses often require cost-benefit analyses of agricultural production systems.

The concept of opportunity costs (OCs) is most frequently used to express the costs of additional watershed service provision. In the context of watershed services, OCs represent any benefits foregone by a farmer upon converting from their current land use to an alternative form of land use that is more watershed service-friendly.

Ideally, PWS are designed in such a way that payments compensate for at least the OCs of additional service provision or the foregone benefits of the land use promoted in order to provide the service. The extent to which such payments are or are not appropriate depends on the alternative land uses in each given area (Pagiola et al. [2010]).

The Demand for Watershed Services

When dealing with demand, we refer to those who are currently benefiting from the delivery of watershed services and to the resources available for protecting and conserving these services (Guedes and Seehusen [2011]).

Besides PWS, which are carried out by national states in Latin America, a user's fee system can be an effective approach to efficiently use water resources. The watersheds of the rivers Piracicaba, Capivari and Jundiaí (PCJ) in the state of São Paulo have implemented such a user's fee system with around 8.8 million of beneficiaries of the Cantareira system (Veiga and Galvadao [2011]). In this situation, an Inter-Municipal Basin Committee was formed to manage a watershed protection fund and contributions to the fund come from the municipal water utility budgets.

As a further alternative, some programmes in the Atlantic Forest region are subsidised by the government, for example: "Bolsa Verde", "ProdutorES de Água" and "Mina D'água" in the states of Minas Gerais, Espírito Santo and São Paulo, which pay rural producers for conservation activities on their properties (Guedes and Seehusen [2011]). These include the protection or restoration of native vegetation areas with a focus on headwaters and riparian forests (Guedes and Seehusen[2011]; Veiga and Galvadao [2011]).

Another significant source of finance for PWS in developing countries comes from the international public sector funding or development assistance (Porras et al. [2008]). A key provider to this funding is the Global Environmental Facility (GEF), which acts as buyer on behalf of service users for conserving global public services. Around 108 million USD and 52 million USD have been made available as World Bank (WB) loans and GEF grants respectively for WB/GEF-supported PWS projects (FAO [2007]). For example, the World Bank has given loans to support the development and implementation of well-known PES programmes in Mexico and Costa Rica (FAO [2007]). Involvement of the private sector in paying for ecosystem services, including in the context of corporate social responsibility, is growing (TEEB [2010c]).

Considering these factors, supply and demand analysis of watershed services provides essential information to assess the economic and financial viability of PWS schemes (IIED [2012a]). Figure 1illustrates our approach in identifying the economic preconditions for a potential PWS scheme in the Guapi-Macacu watershed.

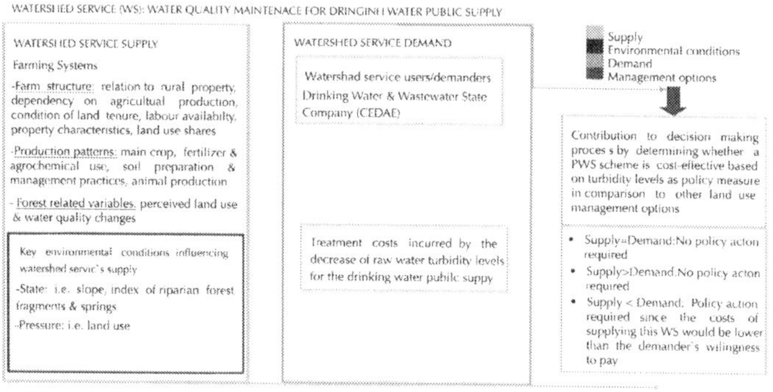

Figure 1: Conceptual framework for assessing watershed services. This framework considers water quality maintenance for public supply considering supply is at the farming system level and demand is at the watershed level.

As the figure suggests, we expect PWS to be a viable policy option only if the willingness of water users to pay (i.e. the demand of the water company) exceeds the opportunity costs (OCs) of additional watershed service provision (supply) incurred by land users in the watershed.

METHODS

The Guapi-Macacu Watershed

The study area lies within the Guapi-Macacu watershed (1,263 km^2), which is located in the Serra do Mar biogeographical region (Ribeiro et al. [2009]) and is a priority conservation target within the Atlantic Forest biodiversity hotspot in the state of Rio de Janeiro (CEPF [2001]), see Figure 2.

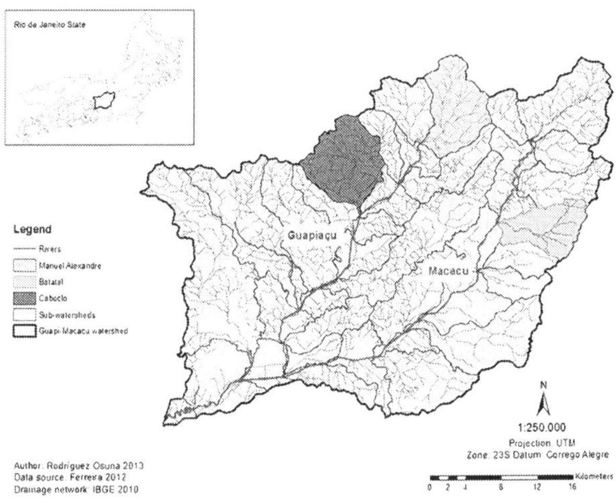

Figure 2: The Guapi-Macacu watershed within the state of Rio de Janeiro.

The Atlantic Forest biome is characterised by historically high deforestation rates (Dean [1997]), resulting in a highly fragmented area with numerous isolated and disperse forest fragments in a landscape dominated by agricultural production systems (Nehren et al. [2013]; Ribeiro et al.[2009]). This biome supplies 135 million people with water (Pria et al. [2013]); however, a mere 11 to 16 percent of its original forest cover is left (Ribeiro et al. [2009]).

Forest cover in the state of Rio de Janeiro and more specifically in the Guapi-Macacu watershed, is comparatively high due to its rugged topography. Unfortunately, continuous forest cover occurs mostly on steep slopes that are inappropriate for agriculture, while the foothills and lowlands are dominated by agriculture and pasture (Nehren et al. [2013]; Strobel et al. [2007]). In 2008, the land cover/land use of the Guapi-Macacu watershed consisted of forests in all stages (48.8%), pastures (41.4%) and agriculture (4.4%). The remaining 5.4% were covered by urban areas, water bodies, bare soil, rock outcrop, wetlands and mangrove (Fidalgo et al. [2008]).

Rivers and creeks in the Guapi-Macacu watershed originate mainly in the State Park "Três Picos", but some stem from the National Park "Serra dos Órgãos" and in the State Ecological Station "Paraíso". The main rivers Macacu and Guapiaçu originate within the State Park and constitute the main components of this watershed (Strobel et al. [2007]).

The water intake point of the water supply is located in the lower part of the watershed and managed by the Drinking Water and Wastewater State Company (CEDAE). Due to the good water quality coming from the springs of the rivers Macacu and Guapiaçu, several mineral water companies have been established in the region, as well as enterprises for which water is an essential input (Strobel et al. [2007]).

In the Guapi-Macacu watershed, various demands in water supply to domestic, industrial and agricultural consumption are taking place as a result of the good water quality from the water sources of the main rivers (Strobel et al. [2007]). Moreover, the ongoing construction of a new petrochemical complex (COMPERJ) in the downstream area of the watershed is expected to further increase both water demand and pressure on natural resources in the study area (Pedreira et al.[2009]).

Within the watershed, three sub-watersheds (see Figure 3): (I) Manuel Alexandre, (II) Batatal and (III) Caboclo were selected for this study. Each of these sub-watersheds represents the different types of land cover mix found in the area, namely "forest", "agriculture" and "pasture", which account altogether for around 95% of the total watershed area. By selecting three sub-watersheds with greatly varying land use patterns and intensities, we were able to compare them with respect to their differences on agricultural profitability and their distinctive impact on water quality parameters: in this case turbidity levels.

1. The sub-watershed Manuel Alexandre is located in the Ecological Reserve of Guapiaçu (REGUA) and represents a well-preserved landscape with a high proportion of forest land (87%) (Fidalgo et al. [2008]). It therefore served as the reference for a nature-

near, less disturbed forest ecosystem. This area includes a private reserve with low human impact resulting primarily from ecotourism in the form of birdwatchers. Most of the area within this sub-watershed is protected under the REGUA Association, which is financially supported by the Brazilian Atlantic Forest Trust (BART) with the stated objective of protecting the Atlantic Rainforest of the upper Guapiaçu river watershed. This is mainly done by enlarging protected areas through land purchase.

2. In contrast, the sub-watershed Batatal represents a mixed system of the most relevant land uses with a mosaic of forest fragments (69%), pastures (28%) and agriculture (4%) considering the land use classification by Fidalgo et al. ([2008]). Predominantly, banana (perennial) is found in higher elevation areas, followed by annual crops mainly found in flat areas or lowlands (cassava, maize, beans and vegetables). Forest fragments in different stages of ecological succession are found in high elevation areas.
3. In the Caboclo sub-watershed, the predominant land use type is forest (81%), followed by pasture (14%) and agriculture (3%) according to Naegeli ([2010]) and Fidalgo et al. ([2008]). Agricultural systems with a considerably higher intensity than in Batatal are found in this sub-watershed, mainly along the floodplain. The most common annual crops are maize, cassava, beans and vegetables, whereas perennial crops are rare. The higher elevation area is within the boundaries of the State Park Três Picos, where agricultural activities are carried out within the buffer area of this Park.

Figure 3: Sub-watershed landscapes of three rivers: Manuel Alexandre (left), Batatal (upper right) and Caboclo (lower right).

Both sub-watersheds, Batatal and Caboclo have undergone historical exploitation cycles, resulting in high deforestation, forest fragmentation and degradation, as well as intense soil erosion (Nehren et al. [2013]).

Methods to Assess Supply of Watershed Services

To calculate the opportunity costs (OCs) related to the provision of watershed services under varying land use systems, we carried out a cost-benefit analysis of representative farming systems in the region. For this, we developed detailed individual budgets for all activities within a given farming system (for definition see Beets [1990]). Activity budgets summarised revenue and cost information and were finally aggregated to calculate the average rate of return for each land use type (WBI [2011]). Crop budgets were compared for coherence with official current production costs used by the Rural Extension and Technical Assistance Agency (EMATER) of the state of Rio de Janeiro.

Our target population was households practicing some degree of agriculture at the sub-watershed level considering the sub-watersheds of Batatal and Caboclo. A household survey was carried out in two field campaigns in 2011 and 2012 with the permanent support of key local producers, EMATER of the municipality of Cachoeiras de Macacu, the City Council of this municipality and Embrapa Soils scientists. Expert interviews were carried out with the Director of the REGUA Reserve and other staff members in Manuel Alexandre sub-watershed, to better understand land use history and recent management practices in the region.

To define the sample universe, we created an inventory using the indirect census technique following Forero ([2002]). This process consisted of a participatory mapping exercise based on recent aerial imagery provided by the City Council of Cachoeiras de Macacu. This enabled us to assemble a list of all farm units within the sub-watersheds, which was confirmed by extensive field visits and supported thoroughly by local experts including a member of the Agricultural Department of the City Council (Cachoeiras de Macacu), the President of the Faraó Farmer's Association (A.L.A.F.) in the sub-watershed of Batatal, a member of the Rural Workers Union (in the sub-watershed of Caboclo) and the

Director of the REGUA Reserve (Manuel Alexandre sub-watershed). As a result, a total of 32 households in Batatal and 60 in Caboclo were identified, of which 78 households within the two populated sub-watersheds were interviewed using a semi-structured survey. No interviews were made in the reference site of Manuel Alexandre sub-watershed. The sample size obtained is supported by Angelsen et al. ([2011]), who suggest a minimum sample size of 25–30 households from each community. This is valid for communities with 100 to 500 families. The designed questionnaire was based on various scientific publications and reports (see Rodriguez Osuna [2013]; Angelsen et al. [2011]; WBI [2011]; Gaese [2009]; Instituto Terra Mater [2009]; Forero [2002]).

Throughout the course of our fieldwork, two survey rounds were carried out. The first survey round included a random sampling of farm units within each sub-watershed to define "representative farming systems". Important selection criteria for these farming systems as suggested by Zimmer et al. ([2009]) and local experts included mainly: farm size, land tenure, production programme and agricultural management practices, and average location of farms in terms of metres above sea level (m.a.s.l.).

Once such farming systems were defined, a second survey round was launched to explore in-depth characteristics of farming systems with special attention to the inputs and outputs that are relevant to profitability among such systems.

In the sub-watershed of Batatal, we divided the farming systems by location in upland versus lowland, since this division significantly affects production patterns. Farm units in the uplands are located at an average altitude of 344 m.a.s.l. in contrast to those in the lowlands located at ca. 83 m.a.s.l. Agricultural production in the lowlands of Batatal is comparable to those sub-watersheds located along the Macacu River. The same occurs in Caboclo, which is representative for sub-watersheds along the Guapiaçu River (Figure 2).

Subsequently, a cost-benefit analysis was carried out for each farming system. The occurrence of each farming system was estimated and validated through local expert consultation as suggested by Angelsen et al. ([2011]).

Our sampling strategy focused on capturing the diversity of smallholder production systems in the region, yet our total sample size was too small to obtain a representative sub-sample of the large cattle operations that dominate in the lower part of the watershed. For the cost-benefit analysis of cattle production systems, we thus relied on additional in-depth interviews with a group of livestock producers deemed representative by officers of EMATER.

Based on interviews with selected livestock producers and secondary data on livestock systems in this area (see Quintana [2012]), we calculated livestock activity budgets for three slope categories: 1) ≤15°; 2) 16-25° and 3) >25°. These budgets were calculated under the assumption that profits for livestock production decreased with increasing slope, because of lower productivity of pasture, among other factors. This assumption is based on interviews with local farmers.

Once profits for agricultural and livestock systems were obtained, they were extrapolated to the watershed level using a Landsat based land cover classification that identified "agriculture" and "pasture" areas (Pedreira et al. [2009]; Fidalgo et al. [2008]). The agricultural profit calculated in the selected sub-watersheds (only considering lowland areas) was applied to all sub-watersheds in the same river network. In the sub-watershed of Batatal, we divided farming systems located in uplands and lowlands since this division influences significantly production patterns. This was not necessary in the Caboclo sub-watershed, where all farming systems are located in the lowlands.

The profit derived from farming systems is equal to the opportunity cost of converting agricultural or pastoral lands to forest, thereby reducing turbidity. For example low OCs are associated with low profits from current land use. Per hectare OC estimates for each sub-watershed thus represent the weighted average per hectare profits from the respective land cover types.

We relied on a spatial analysis of the vulnerability of water resources in the study area (Ferreira[2012]), which was understood as the likelihood of watershed service loss. In this case, our assessment considered state and pressure indicators following the Driving Forces, Pressure, State, Impact and Response (DPSIR) framework (see Borja et al. [2006]), where 50% were state indicators including: geomorphology, hydrogeology, drainage density, soils, index of circularity, index of areas of permanent protection (APP) fragments and slope. APP areas are established by the Brazilian Forest Code (Federal Law 4771/1965) and are defined as "protected areas, both covered or not with native vegetation, that have the environmental functions of preserving water resources, landscapes, geological stability, biodiversity, and genetic fluxes of flora and fauna, as well as protection of the soil and securing the wellbeing of human populations" (Ministry of Environment [2005]). These areas include a minimum vegetation area to protect riverbanks and headwaters. The other 50% included pressure or anthropogenic indicators such as phosphorous (P) and nitrogen (N) production, land use and road density (Ferreira [2012]). All mentioned factors (state and pressure) were weighed by hydrological expert consultation by Ferreira [2012] (Figure 4). Single indicators were based mostly on published official maps predominantly generated by Embrapa Soils.

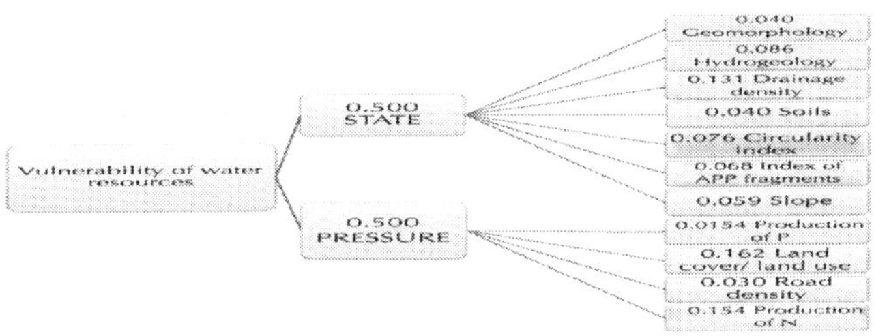

Figure 4: State and Pressure Indicators to assess water resources vulnerability. These indicators were weighed by hydrological expert consultation to assess the vulnerability of water resources in the Guapi-Macacu watershed (Modified from Ferreira[2012]).

Our next step was to identify priority areas for watershed service provision. For this, we overlaid environmental and economic criteria, i.e. vulnerability of water resources and spatial OCs within the Guapi-Macacu watershed. This allowed us to identify the areas where low OCs of shifting land use towards improving watershed services can result in high watershed service payoffs. These sub-watersheds were given the highest priority for intervention with watershed service improvements. Second and third priorities were given to sub-watersheds with high OCs and vulnerability and those with low OCs and vulnerability, respectively. The latter is based on the assumption that environmental goals are more important than cost criteria. However, this could be changed when there are budget restrictions and when there is intent to increase efficiency of payments in compensation schemes for watershed services. The lowest priority areas are those sub-watersheds with high OCs and low vulnerability of water resources.

Methods to Assess the Potential Demand for Watershed Services

Demand for watershed services was assessed by identifying costs related to the end-user of those services (Honey-Roses et al. [2013]). For this particular study case, it included quantifying water treatment costs incurred by the main watershed user, the state water utility (CEDAE). Water treatment costs that may be avoided if forests are restored can be translated as the potential willingness to pay (WTP) for watershed services. For example, if a one unit reduction in turbidity levels implies 10 additional monetary units in treatment costs, the water company's maximum WTP for watershed services will be 10.

We applied the avoided cost method focusing on the annual operational costs of chemical products for the treatment of raw water in the period between 1998 and 2011 from the local state water utility company. This approach required identifying key water quality indicators related to the main operational cost categories of the water utility company.

Based on expert interviews and as suggested by Medeiros et al. ([2011]), Reis ([2004]) and Dearmont et al. ([1998]), we identified turbidity as the key indicator amongst all water quality parameters, since an increase in turbidity implies an increased concentration of suspended solids in surface waters and is likely to reduce the quality of raw water to be treated by the water utility. The conversion of forest to other land uses such as agriculture or pasture caused by farming systems can result in higher turbidity values, which in turn correlate with higher water treatment costs to reduce the concentration of suspended solids for public water supply (Medeiros et al.[2011]; Reis [2004]; Dearmont et al. [1998]).

In addition Medeiros et al. ([2011]) and Reis ([2004]) suggest that chemical products account for close to 60% of annual operational costs in treatment plants in São Paulo (Reis [2004]). These products are used to flocculate the suspended particles, measured as turbidity, that are found in raw water to fulfill regulations on drinking water quality for human consumption (maximum value of 5 NTU).

RESULTS

Vulnerability of Water Resources

Water resource vulnerability is a function of both anthropogenic impact/pressure indicators and of environmental state indicators, so the assessment of vulnerability must account for this set of indicators. When only considering environmental state indicators, sub-watersheds located in higher areas of the Guapi-Macacu watershed tend to have higher vulnerability to anthropogenic pressure than those in the lowlands (Ferreira [2012]). Pressure indicators in a watershed are highly influenced by population density, land use practices, presence of urban settlements, road density and other factors previously mentioned. Therefore, the sub-watersheds with relatively high anthropogenic impacts are those with a high density of urban settlements and rural population nuclei. Sub-watersheds found in higher areas of the watershed have considerably lower impact values (Ferreira [2012]).

Figure 5 shows the sub-watersheds with higher vulnerability in the darker tones and those with lower vulnerability in the lighter tones. The less vulnerable sub-watersheds are found in the lower areas of the watershed and one of these is considered a natural protected area with limitations and restrictions in land use, despite its lower vulnerability.

Figure 5: Vulnerability of water resources in the Guapi-Macacu watershed (Ferreira[2012]).

Agricultural Production and Opportunity Costs

Field surveys carried out in the Batatal and Caboclo sub-watersheds showed differences in production patterns and differences in specific environmental factors that reflect both the effect of farming systems on the provision of watershed services as well as distinct farming systems' profitability.

Relevant characteristics that differ considerably across these two sub-watersheds include land tenure and land use. In Batatal, most of the land is obtained by heritage or with a provisionary land title (66%), followed by banana share-croppers (19%), who are generally entitled to half the harvest. In contrast, in Caboclo most of the population lives within a settlement that keeps a common forest area called a "private reserve". This reserve is protected and is restricted from cultivation. Therefore, 85% of the farm-households in Caboclo are located in the lowlands, while the rest are found on the hillsides. In both sub-watersheds, one farm unit does not surpass 14 hectares, which is considered typical for family-oriented agriculture for the municipality of Cachoeiras de Macacu (MDA [2010]). After considering the specific differences among farming systems in Batatal and Caboclo, four types of farming systems (FS) were classified in Batatal and two in the sub-watershed of Caboclo (Table 3).

Table 3: Summary of annual farming systems profits in the study area

Main crops	Unit	Batatal sub-watershed				Caboclo sub-watershed			Pasture land systems oriented to beef production considering slopea		
		Upland		Lowland		Lowland					
		FS1	**FS2**	**FS3**	**FS4**	**FS5**	**FS6a**	**FS6b**	**≤15**	**16 to 25**	**>25**
Banana	(BRL)	972	972	983							
Cassava			−1,334	3,950	3,950	5,231	5,232	5,232			
Green maize (summer)				2,371	2,371	2,066	2,067	2,067			
Green maize (winter)				3,438	3,438	2,371	2,371	2,371			
Yam				2,861	2,861	6,199	6,199	6,199			
Courgette					3,633						
Gilo								10,941			
Okra							10,092				
Beans (summer)						1,419	1,419	1,419			
Beans (winter)						1,522	1,523	1,523			
Occurrence	(%)	80	20	50	50	70	15	15			

Typical area and crop distribution for each farming system (FS)	(ha)	10.3	10.3 (8.3 banana, 2.0 cassava)	6.0 (2.0 cassava, 2.0 green maize, 1.0 yam, 1.0 banana)	6.0 (2.0 cassava, 2.0 green maize, 1.0 yam, 1.0 courgette)	4.0 (1.0 cassava, 1.5 green maize, 1.0 yam, 0.5 beans)	4.0 (1.0 cassava, 1.5 green maize, 0.5 yam, 0.5 okra or gilo, 0.5 beans)				
Per hectare profit	(BRL ha^{-1})	972	526	3,894	4,336	4,889.7	5,376.3	5,482.5	20	40	100
Average value for extrapolation at the watershed level	(BRL ha^{-1})	Macacu River lowland				Guapiaçú River					
		4,114.8[b]				5,051.6			20	40	100

[a]According to expert interviews, there is a gain of 40 to 60 kg of live animal weight in high slope areas in contrast to those in lowlands with a gain in live animal weight of 120 to 150 kg in this particular watershed; [b]Since upland agriculture in the Batatal sub-watershed (FS1 and FS2) was not detected by the available land use classification (Naegeli [2010]; Pedreira et al. [2009]; Fidalgo et al. [2008]), we were limited to using average per hectare annual profits for lowland agriculture in Batatal (FS3 and FS4) resulting in an average annual per hectare value of 4,114.8 BRL.

Rodríguez Osuna *et al.*

Rodríguez Osuna *et al. Ecological Processes* 2014 **3**:16, doi:10.1186/s13717-014-0016-7

In Batatal, we found that 80% of FS in the uplands are specialised in banana (*Musa sp.*) production (classified as FS 1) and 20% had a mixed system of cassava (*Manihot sp.*) and banana (FS2). In the lowlands of Batatal, two additional FS where agriculture is more intensive were classified as FS3 and FS4 and are equally distributed. FS3 has a production system composed of cassava, green maize (*Zea mays*), yams (*Colocasia sp.*) and courgette (*Cucurbita sp.*), while FS4 has the same mix of cassava, green maize and yams, but banana instead of courgette (Table 3).

Cassava is the dominant crop in the Caboclo sub-watershed, followed by green maize, yams and common beans (*Phaseolus vulgaris*). Cassava is the most cultivated crop, mainly due to low investment requirements and because of relatively stable returns after a cropping period of 8 to 9 months. Green maize requires higher investments; however, it provides relatively rapid returns after only 90 days. Common beans are used to improve soil fertility (nitrogen fixation) and as an alternative to the other products. Other relevant short-cycle products are okra (*Hibiscus esculentus*) and gilo (*Solanum gilo*).

The most common farming system in the Caboclo sub-watershed (FS5) combines cassava, yams, common beans and green maize (70%), while the rest (FS6-FS6a) combine cassava, yams, common beans, green maize and -additionally- okra or gilo (Table 3). Typically, green maize, courgette and beans are planted twice in one cropping cycle.

Agriculture in the uplands of Batatal is clearly less intensive in fertilizer use than in the lowlands, especially given that banana production in the uplands is carried out without fertilization. The remoteness of these locations makes intensive production less attractive than in the lowlands. In the lowlands of Batatal, vegetable producers applied on average of 240 kg of fertilizer per hectare each year. In Caboclo, more than 70% of households used fertilizers for their agricultural production, at an average of 547 kg per hectare each year.

In the uplands of Batatal, all households used herbicides as their basic approach to weed control, whereas a variety of agrochemicals were used in the lowlands. The high use of herbicides for weed control could be the result of the shortage of and high cost of labour that would otherwise undertake this activity. It should also be noted that the Atlantic Forest Law (Law 11.428/1986) and the Brazilian Forest Code (Law

12.651/2012) bring certain limitations to agricultural production. The Atlantic Forest Law bans the conversion of secondary forest into land uses such as agricultural land. As an example, a collapse of the market price for banana around fifteen years ago left many banana plantations uncultivated and secondary forests developed and expanded, leading to abandonment of these plantations. In addition, the Brazilian Forest Code defines for the Atlantic Forest Biome that 20% of rural properties need to be maintained as a permanent forest reserve "Reserva Legal". The Brazilian Forest Code also prohibits the clearing of primary vegetation on steep slopes (>45), along the margins of rivers and streams and in headwater (source) areas, which are classified as areas of permanent protection (APPs) (Ministry of Environment [2005]).

In Batatal, the per hectare average annual profit was estimated at 4,115 BRL and in Caboclo at 5,052 BRL (Table 3). Returns for agriculture tend to be higher in the lowlands, where the intensity of production is higher than in the uplands, and there is a higher use of agricultural inputs especially fertilizers for cash crops.

According to local expert interviews small livestock farmers are considered those with farm size less than 20 hectares, while big-scale producers are considered those with more than 400 hectares. According to official cattle vaccination data in the Municipality of Cachoeiras de Macacu in 2011, an estimation of the herd size can be given on the base of vaccinated animals. This resulted in 27,995 animals in all three districts of this Municipality (Secretary of Agriculture, Cattle Farming, Fisheries and Supply [2011]). Most livestock farmers (90%) have less than 500 animals and small-scale producers are considered in this municipality those with less than 20 animals. Mostly, animals are distributed in paddocks without dividing fences.

Livestock production systems achieved profits of 20, 40 and 100 BRL per hectare annually, depending on slope class (Table 3).

Area-weighted OCs per sub-watershed were spatially mapped, ranging from 14 to 1,660 BRL per hectare (Figure 6). This reflects that extensive pasture is the most important land use in the Guapi-Macacu watershed according to area, which is an activity with comparatively low per hectare profits.

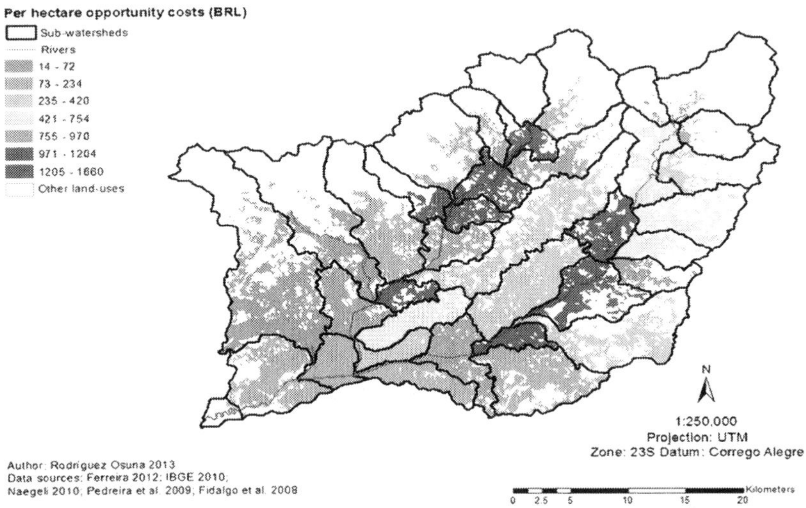

Figure 6: Spatial per hectare opportunity costs in BRL for the land uses "agriculture" and "pasture" within each sub-watershed in the Guapi-Macacu watershed. The white areas within the map correspond to those with land uses other than agriculture in the Guapi-Macacu watershed.

Figure 5 shows that quite a number of high OC areas are located close to the main two river beds, where intensive agriculture predominates. Resulting OCs from agriculture occur only in lowlands and close to the river plain, which have higher nutrient concentration than in higher slope areas.

Low OCs areas are often either located in the steep slope areas of the upper watershed or in the valleys, where extensive pasture areas dominate. However, these areas are also found in the lower parts of the watershed, where a small-scale settlement promoting family-oriented cattle ranching was launched a decade ago by the government of Rio de Janeiro.

Forests predominate mainly in higher slope areas (mostly in white in Figure 6), which originate, to a great extent, in protected areas such as the State Park "Três Picos", the National Park "Serra dos Órgãos" and the State Ecological Station "Paraíso".

Analysis of Environmental and Economic Criteria for Watershed Service Conservation and Improvement

Analysis of environmental (vulnerability of water resources) and economic criteria (OCs of watershed service provision) in the Guapi-Macacu watershed allowed us to identify priority areas for watershed service conservation and improvement. These areas are the land use-based management options (i.e. conversion of pastoral or agricultural lands to forest) with the highest potential of improving water quality and lowest OCs (Figure 7).

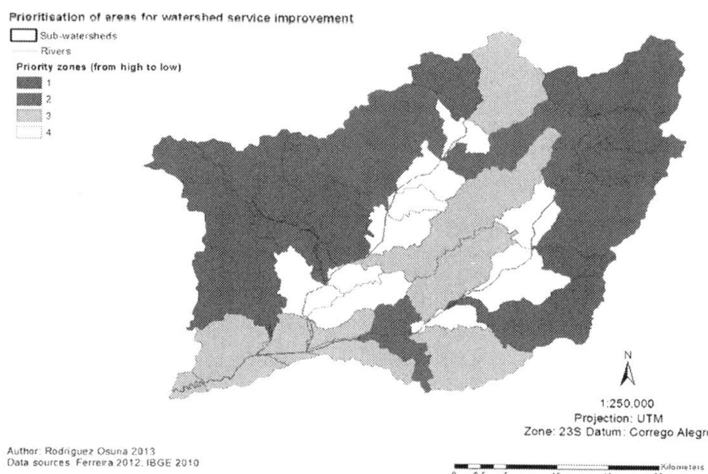

Figure 7: Priority areas for improving or ensuring watershed services of the Guapi-Macacu watershed. Priority areas for improving or ensuring watershed services (water quality) where 1 is considered high and 4 represents low priority.

Our results highlight the sub-watersheds where the vulnerability of water resources are highest and OCs of converting land uses to foster watershed service provision are lowest. The high priority areas (in dark red) are in most cases found in sub-watersheds with steeper average slope levels where impact on water resources was found to be high (Figure 7).

Demand for Watershed Services: Water Treatment

According to micro-economic theory, we interpret the water utility company's demand for chemicals to treat water as its willingness to pay for a desired water quality level (in this case turbidity under 5 NTU for human consumption). The treatment cost of an additional turbidity unit is thus equivalent to the company's potential willingness to pay for any measure that reduces turbidity by the same amount (see avoided cost method, for example, in Perman et al. [2003]).

Based on expert interviews and relevant studies (Medeiros et al. [2011], Reis [2004] and Dearmont et al. [1998]), which identified turbidity as the key water quality parameter relevant for water treatment costs, we applied the avoided cost method to the local water utility company. In Table 4, the main characteristics of the water treatment are presented.

Table 4: Characteristics of water treatment utility in the lower catchment part of the Guapi-Macacu watershed in 2011

Treatment unit: Laranjal - CEDAE	**Municipality: São Gonçalo**
Captivation area from the GMW	1,263 km^2
Forest cover in the watershed	48.8%
Treatment type	Conventional
General treatment phases	Captivation, sedimentation, coagulation, flocculation, decantation, filtration, disinfection, water fluoridation and pH correction
Treated water flow	Average flow in 2011, 5.35 $m^3\ s^{-1}$
Population supplied with treated water	2,000,000 inhabitants
General chemical products used	Aluminium sulphate $Al_2(SO_4)_3$, polyelectrolyte, hexafluorosilicic acid H_2SIF_6, chlorine, calcium oxide CaO
Costs with chemical products for the treatment of raw water	2.31×10^6 BRL (year 2011)

Turbidity of raw water, average values	17.10 NTU
Treated water characteristics	Colour, 2.50 uH
	Turbidity, 0.34 NTU
Total costs with chemical products and electricity in a year for the water treatment unit	For the treatment of 6 $m^3 s^{-1}$,
	• 300,000 [BRL] (monthly expenditure for chemical products)
	• 100,000 [BRL] (monthly expenditure for electricity in the captivation and production area within the unit)

Rodríguez Osuna *et al.*

Rodríguez Osuna *et al.* *Ecological Processes* 2014 **3**:16, doi:10.1186/s13717-014-0016-7

Based on data from CEDAE (treatment unit Laranjal) for the period between 1998 and 2011, we estimated the avoided costs of a reduction of 1% turbidity at 15,510 BRL, considering an average volume of treated water of 174,545 cubic metres at an average cost of 22.2 BRL per 1,000 cubic metres (Table 5). The Pearson correlation coefficient for average annual turbidity levels and costs for chemical products in this water utility for the period between 1998 and 2011 was 0.4. This correlation value is relatively low compared to other studies carried out in Brazil. For example, Reis ([2004]) calculated a correlation value of 0.7 for seven water treatment units in São Paulo and Cabral de Sousa ([2011]) found a coefficient of 0.9 when comparing 10 different sub-watersheds and their treatment costs in São Paulo.

Table 5: Annual avoided costs from 1998–2011 for the local water utility company

Average quantity ($m^3 yr^{-1}$)	**Average cost (1,000 BRL m^{-3})**	**Total cost (BRL)**	**Avoided cost for 1% of turbidity reduction (BRL)**
174,545.3	22.2	3,877,482.0	15,509.9

Rodríguez Osuna *et al.*

Rodríguez Osuna *et al.* *Ecological Processes* 2014 **3**:16, doi:10.1186/s13717-014-0016-7

Reis ([2004]) found that chemical treatment costs in water utility units on the Piracicaba River are 12.7 times higher than the cost of treating water from the Cantareira system. This author argues that this finding can be explained by the considerably lower forest cover in the Piracicaba watershed (4.3%) compared to the Cantareira watershed (27.2%).

Therefore, the geographical location of forests, as well as land use, soil type, geomorphology and predominant geology are considered relevant factors that influence water quality from headwaters (springs) and water treatment costs for public supply. Nevertheless, Reis ([2004]) showed that the percentage of forest cover is often a sufficiently informative indicator of watershed health and thus water quality.

Water quality monitoring obtained in seven monitoring campaigns in the years 2010 and 2011 along the whole watershed demonstrated lower turbidity levels in the highest parts of the sub-watershed, where forest cover is higher. For example, our reference sub-watershed Manuel Alexandre presented an average value of 0.8 NTU at the outlet as compared to 17.4 NTU at a lower region close to the water intake point of the water utility (Paiva et al. [2011]).

Is Paying for Land use Changes that Foster Watershed Services Cheaper than Treating Water?

The estimated OCs of converting land use/land cover to the benefit of water quality can range between 4,000 to 5,000 BRL per hectare each year for agricultural systems and less than 100 BRL for pasture land. However, the actual OCs per hectare in many sub-watersheds are likely to be much lower, especially if land close to rivers and headwaters is covered by extensive pastures. On the demand side, we estimate that the water company CEDAE's maximum WTP for land use change in the watershed based on avoided costs is 15,510 BRL per additional 1% reduction in turbidity levels at the water intake point of the water utility.

Ideally, a full-scale hydrological model for the watershed would provide us with the potential effect of alternative land use scenarios on turbidity at the water intake point of the water utility. In the absence of such a model, we can only provide an informed estimation with regard to the viability of PWS in the watershed. At the high OCs end, it is clearly unrealistic to expect that the conversion of less than 3 hectares of intensively used cropland in the whole watershed (the avoided cost of water quality reduction: 15,510 BRL / maximum per hectare OC: 5,482.5 BRL per ha = 2.83 ha) will result in a 1% reduction of turbidity levels. In contrast, carefully selected pasture and low intensity agricultural sites could potentially be converted into forest (162 to 814 hectares) in the case of pastures. Land use changes at that scale are more likely to bring about measurable changes in turbidity levels if located in zones with a large impact on river water quality.

Unfortunately, given the spatial distribution of pastures and high intensity agriculture in the watershed, there are likely to be limited opportunities to convert large tracts of land at low costs. Payments for forest recuperation may thus likely remain a complementary watershed management measure in our study area. As an alternative to full scale conversion, some simple pasture management techniques, such as limiting access of cattle to the riverbed in lowland pastures (see dark red areas in the centre of Figure 7), could prove to be comparatively low cost and highly effective measures to reduce turbidity levels close to the water intake point of CEDAE.

Inadequate livestock grazing practices can compromise water quality to the point where is considered degraded and highly polluted and not able to meet water quality standards (EPA[2013]). Therefore, excluding livestock from streams and improving range management practices can contribute to reduce turbidity on streams (EPA [2013]).

Although the water supply company's WTP does not match the estimated OCs, payments from other water users are an additional option that can be taken into account in the design of a PWS scheme in this watershed. Particularly important would be the demand of water by COMPERJ (the Rio de Janeiro petrochemical complex). In addition, this assessment was solely carried out on the basis of water quality improvements in terms of turbidity levels, since it was found this service to be relevant for the demanders downstream. However, accounting for additional ecosystem services provided by forest ecosystems would increase the potential WTP for ecosystem services in this region.

DISCUSSION

This study estimated the costs involved in both supply and demand of water quality maintenance and improvement; where we identified priority areas for supply in order to target watershed management measures or support the launch of compensation schemes such as PES or PWS. We estimated demand by assessing the willingness to pay for a water quality improvement in terms of turbidity by the main watershed user, the state water utility company (CEDAE). This approach showed to be helpful for the design of watershed payment schemes in other local contexts (Pagiola et al. [2010], Martinez de Anguita et al. [2011], Garcia-Nieto et al. [2013], Martin-Ortega et al.[2012]). Adjusting off-the-shelf modelling packages, such as SWAT, for our purposes (see for example, Martinez de Anguita et al. [2011] and Quintero et al. [2009]), was deemed inappropriate by local hydrologists. Attributing water quality dynamics to land use changes is one of the most challenging issues in ecosystem services research (De Groot et al. [2010]). Since we lack an appropriate hydrological model, we have left the quantification of the effects of specific land use changes on water quality improvement in our study area for future research.

Some limitations of this study are related to the use of turbidity as a water quality indicator. Land use is most likely the most relevant factor that influence turbidity changes in streams (ECI[2014]). However, turbidity is subject to natural physical and biological variations even if the watershed is forested. The physical factors relate particularly to erosion processes, landslides, and mudslides after heavy rainfalls. These accelerated slope processes that are associated with high sediment loads and turbidity can even take place under closed forest cover, as seen for example during the 2011 mudslides and floods (Nehren et al. [2014]). The nature of soils and geology in the watershed (e.g. erosion of the riverbed) also determines how easily erosion might occur (ECI[2014]).

We found a lower correlation between water treatment costs and turbidity (0.4) than the 0.7 found by Reis ([2004]) in her analysis of 7 treatment utility plants in São Paulo. The use of average values for turbidity and the costs of chemicals used for treatment would most likely mask a higher correlation. However, our correlation is relevant since it supports our conclusion that the WTP from the water supply company is likely to be rather low. Yet, turbidity levels monitored at the water intake point of the water supply facility do not exceed 50 NTU (Paiva et al. [2011]).

Comparing these turbidity levels for water treatment with other cases in Brazil, we found mean turbidity levels of 16 NTU and 64 NTU in São Paulo (the first value between April to October and November to March respectively) at the water intake point of the River Piracicaba. In this particular case, it was found that this treatment plant had the highest costs related to the use of chemical products for water treatment, which in addition might have been correlated with the low forest cover of 4.3% (Reis [2004]). During these same periods mean turbidity in the Cantareira system (dam) varied from 3–9 NTU respectively with the lowest costs for chemical products with a more elevated forest cover 27.1% (Reis [2004]).

However, with regards to the use of chemical products for water treatment and their relative costs, it should be noted that not all chemical products are used to tackle turbidity. The main product used to flocculate suspended material and reduce turbidity is aluminum sulphate, which has the highest share of the cost of chemicals for the treatment process. The use of the other single chemical products is also correlated to the performance of the treatment process, where, for example the better the quality of the final product (i.e. treated/filtered water), the lower the cost with the rest of the chemical products, namely hexafluorosilicic acid, chlorine and calcium oxide.

Agricultural land use clearly represents the most important entry point for watershed managers in the studied area, which is why we focused on quantifying the OCs of changing rural land cover. It should be noted that Lorz et al. ([2011]) demonstrated that urban settlements also negatively impact water quality in terms of turbidity, ammonium (NH_4^+) and Chemical Oxygen Demand (COD). Therefore, future studies should also consider the costs of changes to waste water treatment in upstream urban settlements.

Moreover, we note that the land cover classification used in our analysis does not distinguish between agricultural production systems and underestimates the total land under production due to low resolution (Pedreira et al. [2009]; Fidalgo et al. [2008]). Land use classifications with higher resolution (based on SPOT 5) only exist for the upper part of the Guapi-Macacu watershed, but such data would permit our OC analyses to be extrapolated in more detail.

Comparing our OC estimates to actual transfers in existing PWS schemes in the Atlantic Forest, we find that per hectare payment values range from 10 BRL annually to 566 BRL monthly (Veiga and Galvadao [2011]). These include annual per hectare payments of 176 BRL in the municipality of Extrema, state of Minas Gerais; 25–125 BRL in the watershed of PCJ in the state of São Paulo; 10–60 BRL in the watershed of Guandu, state of Rio de Janeiro; 80–340 BRL in the watershed Benevente, state of Espírito Santo; 80–340 BRL in the Guandu watershed, states of Espírito Santo and Rio de Janeiro; 75–563 BRL paid by the OASIS Foundation, states of Espírito Santo and Paraná; and 175–577 BRL monthly payments in the state of Santa Catarina (Veiga and Galvadao[2011]).

At the regional level in Latin America, we can compare our estimates to well-known schemes, such as the Costa Rican PES programme (FAO [2007]). Here, annual per hectare payments vary from 98 BRL for natural regeneration to 152 BRL for forest preservation and more than 2,332 BRL for new forest plantations for a time period over five years (Ecosystem Marketplace [2010]). The National Programme for Hydrological Environmental Services in Mexico (PSAH) pays between 57–90 BRL per hectare annually depending on the forest type considering the OCs of land conversion (Ecosystem Marketplace [2010]; Muñoz-Piña et al. [2008]). The Ecuadorian PWS Programme in Pinampiro pays landowners around 14–29 BRL per hectare annually (Wunder and Alban [2008]), while the Los Negros programme in Bolivia pays in-kind with beehive boxes for honey production (Asquith and Wunder [2008]). Agricultural OCs in our study area thus clearly exceed the average annual payment in existing PWS schemes by more than a factor of 10.

However, we note that payments made under the above mentioned schemes may focus in other actions other than conversion from agriculture or pastoral lands into forest. In the case of the Atlantic Forest schemes, mostly payments are directed to protect headwater areas and restore and conserve riparian forests. However, actions other than conversion from agriculture or pastoral lands into forest can be paid for (eg. soil conservation practices). The price paid to farmers in the PSAH Mexican scheme has been decided based on the assumption that corn production would be an alternative to conservation

(Ecosystem Marketplace [2010]). The Pinampiro PWS programme pays for forest and páramo protection and regeneration (Wunder and Alban [2008]), while the Los Negros programme compensates for avoided deforestation of cloud forests (IIED [2012b]).

Our results may prove useful for ongoing watershed conservation initiatives, such as the World Bank funded Rio Rural carried out by the State Secretary for Agriculture and Livestock in the state of Rio de Janeiro (RioRural [2013]). This and many other initiatives in the Atlantic Forest region have shown great interest in incentive-based watershed management approaches. Furthermore, Brazilian legislation has been supporting such approaches, such as the Brazilian National Law on Water Resources (Law 9433/1997), which allows for example the establishment of watershed service markets and permits charging for water use as it is taking place in some watersheds in Rio de Janeiro (Veiga [2008]).

Equally relevant is the more recent Law 9985/2000 "SNUC" that adopts the protector-receiver principle, which allows for rewards in exchange for good natural resource stewardship (Strobel et al. [2007]). Funding for incentive-based watershed management may be derived from Brazil's "ecological" tax system, which compensates municipalities for conservation costs based on the value-added tax (ICMS) (Marques [2009]; Veiga [2008]).

In addition, the decree No. 42029/2011 established the Payment for Environmental Services mechanism under the Rio de Janeiro State Programme for Conservation and Revitalization of Water Resources (PROHIDRO). This decree has given legal status to the PES Programmes already underway in the state of Rio de Janeiro, such as the Lagos São João Committee through Good Practices Fund (Funboas) and Guandu Committee through Water and Forest Programme implemented in the municipality of Rio Claro.

This study can potentially contribute to implementation of these laws in the Guapi-Macacu watershed in a more effective way with respect to social, economic and environmental aspects. Given that our study results take into account these different aspects and are based on primary data obtained in the watershed, our study can contribute by identifying limitations and opportunities related to economic watershed compensation programmes.

CONCLUSIONS

Assessing the economic scope of incentive-based watershed management requires knowledge of both the costs of providing additional watershed services through land use and land cover change (service supply) and the willingness to pay (demand) for such services. This study has quantified these two necessary inputs to enable an informed decision-making process in the context of the Guapi-Macacu watershed in the state of Rio de Janeiro, Brazil.

With regard to supply, we found that land users' OCs range between 972–5,482 BRL per hectare for cropland and <100 BRL per hectare for pastures. At the sub-watershed level, zones that use a high proportion of pasture but low proportion of cropland in the land use mix will accordingly have low area-weighted average OCs and vice versa. With regard to demand, we assessed water treatment costs and found a 1% reduction in water turbidity levels to be worth 15,510 BRL on average to the water company, CEDAE.

In absolute terms, only a relatively small share of land in the watershed is cropland (5,560 hectares); the area covered by pasture (52,374 ha) and forest (61,665 ha) is much larger. However, cropland covers a considerable amount of land with potential to provide watershed services and, as a result, high OCs become a critical limitation on strategies that aim to change land use for the benefit of water quality.

Nevertheless, given the spatial distribution of pastures and high intensity agriculture in the watershed, we find that payments for forest recuperation are likely to be cost-effective in sub-watersheds with a high proportion of pasture land. Some of these sub-watersheds also happen to exhibit high levels of water resources vulnerability and thus represent potential priority intervention zones for watershed management (Figure 6). In these areas, watershed planners may reap considerable benefits from active pasture management options, such as silvopastoral systems or strategic fencing of riparian areas. In addition, sustainable agricultural and soil conservation practices can bring additional benefits in terms of reducing and avoiding sedimentation, erosion, which in turn result in improving water quality.

To the extent that smallholders in the study area also depend on subsistence production for their livelihoods, land use based watershed management strategies will have to also consider food security issues, especially in the upper part of the watershed.

Our study area reflects watershed management issues that are typical across many parts of the Atlantic Forest region, which often have diverse agricultural mosaics and thus highly variable OCs. Wherever intensively used cropland dominates in the vulnerable zones of large watersheds, land use planners will find it difficult to rely on the use of payments for watershed services alone. Effective watershed management will then have to be combined with enhanced monitoring and enforcement activities to ensure that the Brazilian Forest Law is complied with, particularly with regard to riparian forests. Research designed with a bottom-up approach that quantifies the potential compliance costs of land users can clearly help policy makers to target both incentives and disincentives in a cost-effective manner.

AUTHORS' CONTRIBUTIONS

VRO undertook the field work, carried out the analysis and wrote the manuscript under her PhD thesis at the University of Leipzig. JB and RBP supported the conception of the research design and data analysis. JB, RBP and UN reviewed and commented on various versions on this manuscript. UN provided insights on ecosystem management aspects. HG and JH contributed to the overall research design and gave valuable comments on aspects related to farming systems analysis and landscape ecology (respectively). All authors read and approved the final manuscript.

ACKNOWLEDGEMENTS

The authors acknowledge financial support of the IPSWAT Programme (financed from the German Federal Ministry of Research-BMBF) and the DINARIO Project: "Climate Change, Landscape dynamics, Land use and Natural Resources in the Atlantic Forest of Rio de Janeiro". We thank the support of the Team of Researchers of MP2 Embrapa Soils of Rio de Janeiro and DINARIO. In our study region, we specially thank Lenilson Biazatti (from the Rural Workers Union), Nicholas Locke (REGUA Association Director), Thabta Matos de Mata from the City

Council of Cachoeiras de Macacu, Jocemir Da Silva (Chief technical extension officer of EMATER in Cachoeiras de Macacu), Demerval Pereira de Sousa (President of A.L.A.F.), technical staff from the state water utility company (CEDAE – Imunana Laranjal) and all farmers who enriched this study with their valuable time. We are also very grateful for the proofreading of this text by Dr. Kylie Quinn and Meghan Doiron.

REFERENCES

1. Angelsen A, Larsen HO, Lund J (2011) Measuring livelihoods and environmental dependence: Methods for research and fieldwork. Routledge, London.
2. Arriagada RA, Ferraro PJ, Sills EO, Pattanayak SK, Cordero-Sancho S (2012) Do payments for environmental services affect forest cover? A farm-level evaluation from Costa Rica. Land Econ 88(2):382-399
3. Asquith N, Wunder S (2008) Payments for Watershed Services. The Bellagio Conversations. Fundación Natura Bolivia, Bolivia.
4. Balvanera P, Uriarte M, Almeida-Lenero L, Altesor A, DeClerck F, Gardner T, Hall J, Lara A, Laterra P, Pena-Claros M, Silva Matos DM, Vogl AL, Romero-Duque LP, Arreola LF, Caro-Borrero AP, Gallego F, Jain M, Little C, de Oliveira XR, Paruelo JM, Peinado JE, Poorter L, Ascarrunz N, Correa F, Cunha-Santino MB, Hernandez-Sanchez PA, Vallejos M (2012) Ecosystem services research in Latin America: The state of the art. Ecosystem Services 2:56-70
5. (1998) Improving Water Utilization from a Catchment Perspective. SWIM Paper. International Water Management Institute, Colombo, Sri Lanka.
6. Beets WC (1990) Raising and Sustaining Productivity of Smallholder Farming Systems in the Tropics. A Handbook of Sustainable Agricultural Development. A Handbook of Sustainable Agricultural Development. AgBe Publishing, Holland.
7. Bennett G, Carroll N, Hamilton K (2013) Charting New Waters. State of Watershed Payments. Forest Trends, Washington.
8. [http:/ / www.springer.com/ environment/ environmental+management/ book/ 978-94-007-5175-0]

Boerner J, Vosti S (2012) Managing Tropical Forest Ecosystem Services: An Overview of Options. In: Muradian R, Rival L (eds) Governing the Provision of Ecosystem Service, vol 4. Springer, pp 21–46,

9. Borja A, Galparsoro I, Solaun O, Muxika I, Tello EM, Uriarte A, Valencia V (2006) The European Water Framework Directive and the DPSIR. A methodological approach to assess the risk of failing to achieve good ecological status. Estuar Coast Shelf Sci 66(1):84-96
10. Calder I (2005) Blue revolution: Integrated land and water resources management 2nd edition. Earthscan, London
11. (2001) Perfil do Ecossistema Mata Atlantica, Hotspot de Biodiversidade – Brazil. Critical Ecosystem Partnership Fund,
12. Costanza R, d'Arge R, de Groot R, Farber S, Grasso M, Hannon B, Limburg K, Naeem S, O'Neill RV, Paruelo J, Raskin RG, Sutton P, van den Belt M (1997) The value of the world's ecosystem services and natural capital. Nature 387(1):253-260
13. Daily GC, Matson PA (2008) Ecosystem services: From theory to implementation. Proc Natl Acad Sci 105(28):9455-9456
14. de Cabral Sousa WJ (2011) Pagamento por Servicos Ecossistemicos: Mata Ciliar, Erosao, Turbidez e Qualidade de Agua: Produtos Tecnicos 1. Projeto de Recuperacao de Matas Ciliares, Brazil.
15. De Groot RS, Wilson MA, Boumans RM (2002) A typology for the classification, description and valuation of ecosystem functions, goods and services. Ecol Econ 41(3):393-408
16. De Groot RS, Alkemade R, Braat L, Hein L, Willemen L (2010) Challenges in integrating the concept of ecosystem services and values in landscape planning, management and decision making. Ecol Complex 7(3):260-272
17. Dean W (1997) With Broadax and Firebrand: The Destruction of the Brazilian Atlantic Coastal Forest. University of California Press, United States of America.
18. Dearmont D, McCarl BA, Tolman DA (1998) Costs of water treatment due to diminished water quality: a case study in Texas. Water Resour Res 34(4):849-853
19. [http://assets.panda.org/downloads/runningpurereport.pdf] Dudley N, Stolton S (2003) Running Pure: The Importance of Forest

Protected Areas to Drinking Water, The Arguments for Protection Series. World Bank/WWF Alliance for Forest Conservation and Sustainable Use, United Kingdom, . Accessed 10 Jan 2011

20. (2014) Chapter 3 – Streams. In: Total Suspended Solids and Turbidity in Streams. Department of Ecology, State Washington. [http://www.ecy.wa.gov/programs/wq/plants/management/joysmanual/streamtss.html] http://www.ecy.wa.gov/programs/wq/plants/management/joysmanual/streamtss.html accessed 7 may 2014
21. (2010) Water Market: Mexico Payment for Hydrological Services. Ecosystem Marketplace: A Forest Trends Initiative. The Katoomba Group.
22. Engel S, Pagiola S, Wunder S (2008) Designing payments for environmental services in theory and practice: An overview of the issues. Ecol Econ 65(4):663-674
23. [http://water.epa.gov/lawsregs/lawsguidance/cwa/tmdl/nutriosocreek.cfm] EPA (2013) Total Maximum Daily Loads (TMDLs) at Work: Arizona. Committed Landowner Implements and Advocates Improved Grazing Practices in Nutrioso Creek, Reducing Turbidity Levels to Attain Water Quality Standards. United States Environmental Protection Agency, . Accessed 5 May 2014
24. (2007) The State of Food and Agriculture: Paying Farmers for Environmental Services. Food and Agriculture Organization of the United Nations, Rome.
25. Farber S, Costanza R, Childers DL, Erickson J, Gross K, Grove M, Hopkinson CS, Kahn J, Pincetl S, Troy A (2006) Linking ecology and economics for ecosystem management. Bioscience 56(2):121-133
26. Ferreira CEG (2012) Sistema de Suporte a Decisao Espacial Aplicado a Analise da Vulnerabilidade dos Recursos Hidricos na Bacia Guapi-Macacu. Universidade Estadual do Rio de Janeiro, Rio de Janeiro, Brazil.
27. Fidalgo E, Pedreira B, Abreu MD, Moura ID, Godoy M (2008) Uso e Cobertura da Terra da Bacia Hidrografica do rio Guapi-Macacu. Embrapa Solos, Brazil.
28. Forero FA (2002) Sistemas de producción rurales en la Región Andina colombiana: análisis de su viabilidad económica,

ambiental y cultural. Grupo Sistemas de Producción y Conservación/ Instituto de Estudios Rurales (IER), Bogota, Colombia.

29. Gaese H (2009) Demands for Interdisciplinary Research in Human-Ecological Systems. In: Gaese H, Torrico JC, Wesenberg J, Schluter S (eds) Biodiversity and Land Use Systems in the Fragmented Mata Atlantica of Rio de Janeiro, Cuvillier Verlag, Gottingen, Germany.
30. Garcia-Nieto AP, Garcia-Llorente M, Iniesta-Arandia I, Martin-Lopez B (2013) Mapping forest ecosystem services: From providing units to beneficiaries. Ecosystem Services 4(Special Issue on Mapping and Modelling Ecosystem Services):126-138
31. Grieg-Gran M, Porras I, Wunder S (2005) How can market mechanisms for forest environmental services help the poor? Preliminary lessons from Latin America. World Dev 33(9):1511-1527
32. Guedes FB, Seehusen SE (2011) Pagamentos por Servicos Ambientais na Mata Atlanica: licoes aprendidas e desafios. Ministerio do Meio Ambiente-MMA, Brasilia, Brazil.
33. Haines-Young R, Potschin M (2013) Proposal for a Common International Classification of Ecosystem Goods and Services (CICES) for Integrated Environmental and Economic Accounting.
34. [http://www.wri.org/publication/forests-water] Hanson C, Talberth J, Yonavjak L (2011) Forests for Water: Exploring Payments for Watershed Services in the U.S. South. World Resources Institute, WRI Issue Brief. Available via WRI. . Accessed 04 Jul 2014
35. [http:/ / www.environmentalmanager.org/ wp-content/ uploads/ 2008/ 04/ valuation%2520of%2520ecosystems.pdf] Hawkins K (2003) Economic Valuation of Ecosystem Services. University of Minnesota, . Accessed 04 Jul 2014
36. Holmes TP (1988) The offsite impact of soil erosion on the water treatment industry. Land Econ 64(4):356-366
37. Honey-Roses J, Acuna V, Bardina M, Brozovi N, Marce R, Munne A, Sabater S, Termes M, Valero F, Vega L, Schneider DW (2013) Examining the Demand fo Ecosystem Services: The Value of Stream Restoration for Drinking Water Treatment Managers in the

Llobregat River, Spain, vol 90:196.205. Elsevier BV

38. (2012) Paying for Watershed Services: An Effective Tool in the Developing World.
39. (2012) Compensation for Hydrological Environmental Services in Los Negros Cloud Forest. Payments for Watershed Markets-Information on Schemes in Developing Countries.
40. Instituto Terra Mater (2009) Relatório da Análise de Percepção Ambiental e sobre Pagamentos por Serviços Ambientais nas Microbacias do Moinho e Cancan. Produto 3. Serviços de monitoramento sócio-econômico e de percepção ambiental em microbacias piloto – "Projeto de pagamento de serviços ambientais". Contrato no PRMC/GEF/BIRD, Piracicaba, Brazil
41. López A (1997) Análise dos Custos Privados e Sociais da erosão do solo- o caso da bacia do Rio Corumbatai. Escola Superior de Agricultura "Luiz de Queiroz" (ESALQ), University of São Paulo (USP), Piracicaba.
42. Lorz C, Abbt-Braun G, Bakker F, Borges P, Bornick H, Frimmel FH, Gaffron A, Hebben N, Hofer R, Makeschin F, Neder K, Roig HL, Steiniger B, Strauch M, Walde D, Weis H, Worch E, Wummel J (2011) Die Bedeutung von Landnutzungsanderungen fur ein Integriertes Wasserressourcen-Management. Eine Fallstudie aus dem westlichen Zentral-Brasilien. . Fachberichte Wasserversorgung. Fachberichte Wasserversorgung, Vulcan Verlag, Essen, Germany.
43. Marques FM (2009) Valoracao dos Servicos Ambientais da Floresta de Mata Atlantica associados a Qualidade e Quantidade da Agua na APA do Sana. Curso de Pos-Graduacao em Agronomia Ciencia do Solo. Seropedica, Universidade Federal Rural do Rio de Janeiro (UFRRJ), Rio de Janeiro, Brazil.
44. Martinez de Anguita P, Rivera S, Beneitez JM, Cruz F, Espinal FM (2011) A GIS cost-benefit analysis-based methodology to establish a payment for environmental services system in watersheds: application to the Calan River in Honduras. J Sustain For 30(1–2):79-110
45. Martin-Ortega J, Ojea E, Roux C (2012) Payments for Water Ecosystem Services in Latin America: Evidence from Reported

Experience. TEEB Conference 2012. Mainstreaming the Economics of Nature: Challenges for Science and Implementation, Leipzig, Germany. ID143

46. (2010) Programa Nacional de Crédito Fundiário (PNCF). Linha de Financiamento combate à pobreza rural. Ministry of Agrarian Development (MDA). Secretaria de Reordenamento Agrario, Brasilia, Brazil.
47. (2005) The Millenium Ecosystem Assessment. Island Press, Washington D.C.
48. Medeiros R, Young CEF, Pavese HB, Araujo FFS (2011) Contribuição das unidades de conservação para a economia nacional: Sumário Executivo. UNEP-WCMC, Brasilia, Brazil.
49. (2005) Rio Floresta. Assistencia Tecnica e extensao florestal aos agricultores familiares da Mata Atlantica do Estado do Rio de Janeiro. Bases Legais e tecnicas para implantacao de projetos florestais. Ministry of Environment Niteroi, Rio de Janeiro, Brazil.
50. [http://rua.ua.es/dspace/handle/10045/7641?locale=en] Montes C (2007) Del desarrollo sostenible al servicio de los ecosistemas. Asociacion Espanola de Ecologia Terrestre. Ecosistemas Revista Cientifica y Tecnica de Ecologia y Medio Ambiente 16(3). Available via RUA.. Accessed 04 Jul 2014
51. Moore W, McCarl B (1987) Off-site costs of soil erosion: A case study in the Willamette Valley. West J Agric Econ 12(1):42-49
52. Muñoz-Piña C, Guevara A, Torres JM, Braña J (2008) Paying for the hydrological services of Mexico's forests: Analysis, negotiations and results. Ecol Econ 65(4):725-736
53. [http:/ / dinario.fh-koeln.de/ 2011/ F.%20Naegeli-%20 Evaluation%20of%20f orest%20fragmentation%20and%20 land% 20use.pdf] Naegeli F (2010) Evaluation of Forest Fragmentation and Land Use Change Patterns using Remote Sensing Techniques and Field Methods, Master Thesis. Cologne University of Applied Sciences, Abstract available via DINARI. Accessed 04 Jul 2014
54. Nehren U, Kirchner A, Sattler D, Turetta AP, Heinrich J (2013) Impact of natural climate change and historical land use on landscape development in the Atlantic Forest of Rio de Janeiro,

Brazil. An Acad Bras Cienc 85(2):497-518

55. [http://postconflict.unep.ch/publications/DRR_CASE_STUDIES_&_EXERCISES.pdf] Nehren U, Sudmeier-Rieux K, Sandholz S, Estrella M, Lomarda M, Guillén T (eds) (2014) The Ecosystem-based Disaster Risk Reduction Case Study and Exercise Source Book. UNEP/CNRD, . Accessed 04 Jul 2014

56. Pabon-Zamora L, Fauzi A, Halim A, Bezaury-Creel J, Vega-Lopez E, Leon F, Gil L, Cartaya V (2008) Protected Areas and Human Well-Being: Experiences from Indonesia, Mexico, Peru and Venezuela. Secretariat of Convention on Biological Diversity, Montreal.

57. Pagiola S, von Ritter K, Bishop J (2004) Assessing the Economic Value of Ecosystem Conservation. The World Bank Environment Department, Washington, United States of America.

58. Pagiola S, Zhang W, Colom A (2010) Can payments for watershed services help finance biodiversity conservation? A spatial analysis of Highland Guatemala. J Nat Resour Pol Res 2(1):7-24

59. Paiva M, Penedo S, Kuenne A, Prado RB, Schuler AE (2011) Qualidade da Agua e Exportacao de Sedimentos em Sub-bacias dos rios Guapi-Macacu- Bioma Mata Atlantica- RJ. In: XXXIII Congresso Brasileiro de Ciencia do Solo. Solos nos Biomas Brasileiros: Sustentabilidade e Mudancas Climaticas. Embrapa, Uberlandia /Minas Gerais, Brazil. [http://ainfo.cnptia.embrapa.br/digital/bitstream/item/51926/1/2315-1.pdf] Available via Embrapa. http://ainfo.cnptia.embrapa.br/digital/bitstream/item/51926/1/2315-1.pdf. Accessed 04 Jul 2014

60. Pattanayak SK, Wunder S, Ferraro PJ (2010) Show me the money: Do payments supply environmental services in developing countries? Rev Environ Econ Policy 4(2):254-274

61. Pedreira B, Fidalgo ECC, Abreu MB, Epiphanio JCN, Galvao LS (2009) Mapeamento do uso e cobertura da terra da bacia hidrografica do rio Guapi-Macacu, RJ. Anais XIV Simposio Brasileiro de Sensoriamento Remoto. Natal, INPE. pp 2111–2118

62. [http:/ / dutraeconomicus.files.wordpress.com / 2014/ 01/ roger-perman-yue-ma-michael-common- david-maddison-james-mcgilvray-natu ral-resource-and-environmental-econ omics-3rd-edition-2003.pdf] Perman R, Ma Y, McGilvray J, Common M (2003) Natural Resource and Environmental Economics,

Third Editionth edn. Pearson Education Limited, Available via WordPress. . Accessed 04 Jul 2014

63. [http://pubs.iied.org/13542IIED.html] Porras I, Grieg-Gran M, Neves N (2008) All that Glitters: A Review of Payments for Watershed Services in Developing Countries, vol Natural Resource Issues 11. International Institute for Environment and Development, Available via IIED. . Accessed 04 Jul 2014
64. Postel B, Barton H, Jr T (2005) Watershed protection: Capturing the benefits of nature's water supply services. Nat Res Forum 29:98-108
65. Pria DA, Diederichsen A, Klemz C (2013) Pagamento por Servicos Ambientais. Uma estrategia para a conservacao ambiental nas regioes produtivas do Brasil? Sustentabilidade em. Debate 4(1):317.340
66. [http:/ / www.google.de/ url?url=http:/ / comunidadpmpca. uaslp.mx/ documento.aspx%3FidT%3D258&rct=j&fr m=1&q= &esrc=s&sa=U&ei=z9q2U8KDLOTmy wPbw4CgCA&ved=0C CQQFjAD&usg=AFQjCNF ldcTMmPE0Err5jbaXkOJOA3FeJA] Quintana B (2012) Native Tree Species in Silvopastoral Systems: A Bioeconomic Assessment in Cachoeiras de Macacu, RJ-Brazil, Master Thesis. Master of Science awarded by Universidad Autónoma de San Luis Potosí and Cologne University of Applied Sciences. Cologne University of Applied Sciences. Available. . Accessed 04 Jul 2014
67. Quintero M, Wunder S, Estrada R (2009) For services rendered? Modeling hydrology and livelihoods in Andean payments for environmental services schemes. For Ecol Manag 258(9):1871-1880
68. Reis LVS (2004) Cobertura Florestal e Custo do Tratamento de Águas em Bacias Hidrográficas de Abastecimento Público: Caso do Manancial do Município de Piracicaba. University of São Paulo (USP), Piracicaba.
69. Ribeiro MC, Metzger JP, Martensen AC, Ponzoni FJ, Hirota MM (2009) The Brazilian Atlantic Forest: How much is left, and how is the remaining forest distributed? Implications for conservation. Biol Conserv 142(6):1141-1153
70. (2013) Rio Rural. Programa de Desenvolvimento Rural Sustentavel em Microbacias Hidrograficas. Governo do Rio de Janeiro.

Secretaria de Agricultura e Pecuaria, Rio de Janeiro, Brazil.

71. Rodriguez Osuna V (2013) Smallholder Production and Climate Risk: The Lower Amazon Region, Brazil. LAP Lambert Academic Publishing, Germany.
72. [http:/ / www.teebweb.org/ publication/ the-economics-of-ecosystems-and-bio diversity-teeb-for-water-and-wetlan ds/] Russi D, ten Brink P, Farmer A, Badura T, Coates D, Forster J, Kumar R, Davidson N (2013) The Economics of Ecosystems and Biodiversity for Water and Wetlands. Institute for European Environmental Policy (IEEP) & Ramsar Secretariat. Available via TEEB. . Accessed 04 Jul 2014
73. Santos DG, Domingues AF, Gisler CVT (2010) Gestão de recursos hídricos na agricultura: O Programa Produtor de Água. In: Prado RB, Turetta AP, Andrade AG (eds) Manejo e Conservação do Solo e da Água no Contexto das Mudanças Ambientais, Brazil, Rio de Janeiro. pp 353-376
74. (2011) Government of Rio de Janeiro. Campanha de Vacinalção contra a febre aftosa. Cachoeiras de Macacu Region. Núcleo de Defesa Sanitária- Escritorio Cachoeiras Macacu, Rio de Janeiro, Brazil.
75. Stanton T, Echavarria M, Hamilton K, Ott C (2010) State of Watershed Payments. An Emerging Marketplace. Executive Summary. Ecosystem Marketplace: A Forest Trends Initiative.
76. Strobel JS, De Sousa JR WC, da Motta RS, Amend MR, Goncalves DA (2007) Critérios Econômicos para a Aplicação do Princípio do Protetor – Recebedor: Estudo de Caso do Parque Estadual dos Três Picos Serie Tecnica. Conservation Strategy Fund, Brazil.
77. (2009) The Economics of Ecosystems and Biodiversity for National and International Policy Maker. Summary: Responding to the Value of Nature. The Economics of Ecosystems and Biodiversity (TEEB), Wesseling, Germany.
78. (2010) The Economics of Ecosystems and Biodiversity (TEEB). Ecological and Economic Foundations, Earthscan, London and Washington.
79. (2010) Mainstreaming the Economics of Nature: A Synthesis of the Approach, Conclusions and Recommendations of TEEB. The

Economics of Ecosystems and Biodiversity (TEEB), Malta.

80. (2010) TEEB for Business Report. The Economics of Ecosystems and Biodiversity (TEEB), Malta.
81. Thomas JM, Callan SJ (2010) Economia Ambiental: Aplicacoes, Politicas e Teoria. Cencage Learning, São Paulo.
82. van Noordwijk M (2005) RUPES typology of environmental service worthy of reward. ICRAF-Southeast Asia, Bogor.
83. Veiga FC (2008) A Construcao dos Mercados de Servicos Ambientais e suas Implicacoes para o Desenvolvimento Sustentavel no Brasil. UFRRJ, Rio de Janeiro, Brazil.
84. Veiga FC, Galvadao M (2011) Iniciativas de PSA de conservacao dos recursos hidricos na Mata Atlantica. In: Guedes BF, Seehusen ES (eds) Pagamentos por Servicos Ambientais na Mata Atlanica: licoes aprendidas e desafios, d Ministerio do Meio Ambiente, Secretaria de Biodiversidade e Florestas, Departamento de Conservacao da Biodiversidade, Brasilia, Brazil.
85. (2011) Estimating the opportunity costs of REDD+. The World Bank Institute, Washington DC.
86. (2011) Mainstreaming Ecosystem Services Initiative (MESI). World Resources Institute, Washington, United States of America.
87. Wunder S, Alban M (2008) Decentralized payments for environmental services: The cases of Pimampiro and PROFAFOR in Ecuador. Ecol Econ 65(4):685-698
88. Zimmer Y, Deblitz C, Seifert K (2009) Die globale Landwirtschaft besser verstehen : Agrarokonomen betreiben internationales Netzwerk agri benchmark. Agribenchmark, Braunschweig, Germany.

Chapter 7

Policies in Support of Pastoralism and Biodiversity in the Heterogeneous Drylands of East Africa

An MO Notenbaert[1], Jonathan Davies[2], Jan De Leeuw[1], Mohammed Said[1], Mario Herrero[1], Pablo Manzano[2], Michael Waithaka[3], Abdilahi Aboud[4], and Shadrack Omondi[5]

[1]International Livestock Research Institute, Nairobi, Kenya

[2]International Union for Conservation of Nature, PO Box 68200, Nairobi, Kenya

[3]Association for the Strengthening of Agricultural Research in Eastern and Central Africa, 765, Entebbe, Uganda

[4]Department of Natural Resources, Egerton University, PO Box 536, Njoro, Kenya

[5]RECONCILE, PO Box 7150, Nakuru, Kenya

ABSTRACT

Research and practice are increasingly demonstrating the environmental benefits of pastoralism and the opportunity for sustainable development of pastoral communities through a combination of livestock and biodiversity-related business. To take full advantage of the potential biodiversity-pastoral synergies, it will be crucial to put in place supporting policies. They need to be embedded in the context of overall pastoral development. However, rangelands and pastoral societies in drylands are heterogeneous, and development options cannot be assumed to be uniform. Factors such as aridity, access to markets and population pressure influence the constraints and the opportunities for both pastoral and non-pastoral communities. We describe the differential challenges to development along these gradients and identify investment priorities if the policy objectives were to support the complementarities between pastoralism and biodiversity conservation.

BACKGROUND

In recent years, it has become more evident to researchers that livestock mobility produces positive environmental externalities: environmental benefits such as biodiversity conservation and stimulated pasture growth. Understanding and promoting these benefits, and thus feeding back positively on the system, have been shown to make good economic sense (McGahey et al. 2008;Davies et al. 2010). Herd mobility is not only essential for effective risk management, it also enables pastoralists to harness the environmental variability and enhance livestock production (Swallow 1994; Nori 2007; Krätli and Schareika 2010).

Mobility can be driven by many factors, including seasonal availability of (or requirement for) resources, evasion of seasonal stresses such as livestock disease, access to services such as markets and avoidance of conflict.

The ecological case was made most emphatically with the 1993 publication *Rangelands at Disequilibrium* (Behnke et al. 1993). This publication argues that the extreme uncertainty of resource availability in dryland environments demands a mobile, opportunistic management strategy (see also Scoones 1995). The publication also outlined the phenomenon of ecological flux, which provides a compelling argument against perceived rangeland degradation by pastoralists. In*Managing Mobility in African Drylands*, Niamir-Fuller (1999) argues that where mobility and customary institutions remain functional, pastoral rangelands are generally observed to be in good environmental condition.

The ecological case is now being reinforced with research into the intimate relationships between herbivores, including domestic livestock, and their environment. Grazing ungulates and grasslands appear to have co-evolved to the point that they provide mutually beneficial services (Stebbins 1987), with herbivore action promoting growth of grasses, dispersal and fecundity of seeds, and overall nutrient and water cycling in rangeland ecosystems. However, the periodicity of grazing and its intensity, rather than cumulative grazing pressure, appears to be the most important factor determining sustainability of grazing. Research appears to show what pastoralists have long argued: that grasslands can suffer from long-term under-grazing as much as from overgrazing, but periodic bursts of intense grazing pressure can be beneficial (Savory 1999; Briske et al. 2008;Vetter 2005). Other research has shown that pastoralists have an intimate relationship with their environment and a rich knowledge that enables them to both protect and exploit their environment and conserve the rangelands' biodiversity on which they depend (McGahey et al. 2008; Aboud et al. 1996).

In summary, well-managed extensive livestock production on communal land is one of the most appropriate uses of the drylands in Africa (Behnke et al. 1993; Scoones 1993; ILRI 2006; UNDP 2006; Neely et al. 2009), both from an ecological and an economic standpoint. The evidence that pastoralism plays an important role in biodiversity conservation is persuasive, and the opportunities for pastoralists to benefit from conservation are also convincing (Norton-Griffiths and Said 2010; FAO 2009). However, rangelands and the pastoral societies making use of them are heterogeneous, and development options thus cannot be assumed to be uniform. Using spatial information, socio-economic analysis and a review of current policy, this paper identifies investment priorities for different dryland zones in Eastern Africa if the policy objective were to support the mutual benefits between pastoralism and biodiversity. We hereby focus into the dryland areas with an aridity index below 0.65 in Ethiopia, Kenya and Tanzania which include arid, semi-arid as well as sub-humid zones (Figure 1). We use the term pastoralism to refer to extensive production of herbivorous livestock using pasture (or browse) in which herd mobility is a central management strategy. In line with the Secretariat of the Convention on Biological diversity (2005), we refer to biodiversity as the variability among all living organisms found in the drylands. It includes diversity within species, between species and of ecosystems in which they live. Wildlife are but part of the species diversity that contribute to biodiversity.

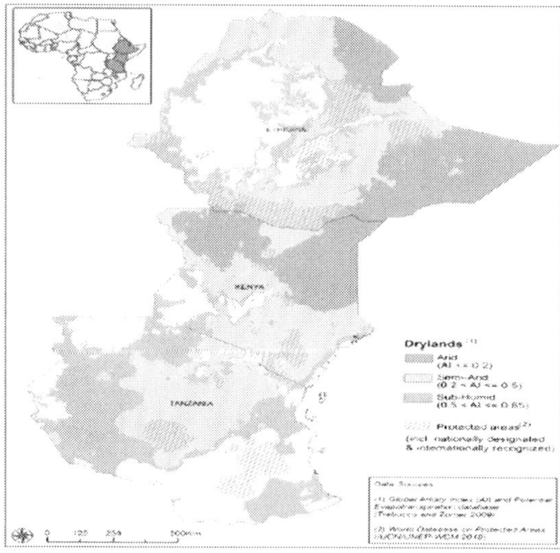

Figure 1: The drylands of East Africa.

The drylands in East Africa contribute considerably to national economies and to society, as they support agriculture, livestock rearing, tourism and wild resource harvesting, and play a critical role in ensuring national food sufficiency (Nassef et al. 2009). The bulk of the meat, milk and other livestock products consumed in the Horn of Africa region comes from the drylands (Kirkbride and Grahn 2008). At the same time, these drylands are home to many unique plant and animal species, which are adapted to the seasonality and extreme unpredictability of precipitation. These adaptations enable them to provide important environmental services such as the conservation of land, water and biodiversity (Mortimore 2005). Some species are also endowed with natural chemicals, for example, the resins (gum arabic) excreted by the acacia tree species that are used in industries. As a result, drylands can provide a range of goods and services, many of which are not effectively measured or sustainably or optimally exploited (Davies and Hatfield 2008). For example, in Uganda, natural products from savannah woodlands and bushlands contributed up to US\$30 ha^{-1} per year^{-1} to the economic well-being of households (Bush et al. 2004). The drylands also support a great diversity of large mammals that have significant global and local cultural value. According to (UNEP 2007), Ethiopia contains 277 mammal species, 626 bird species and 6,500 plant species. (UNEP 2007) estimates mammal diversity in Tanzania and Kenya to be even higher, with 364 and 400 different species, respectively. Figure 2 shows the mammal richness in Eastern Africa as extracted from the African Mammal Databank (Boitani et al. 1999). It indicates high mammal species richness both in the drylands and the humid areas.

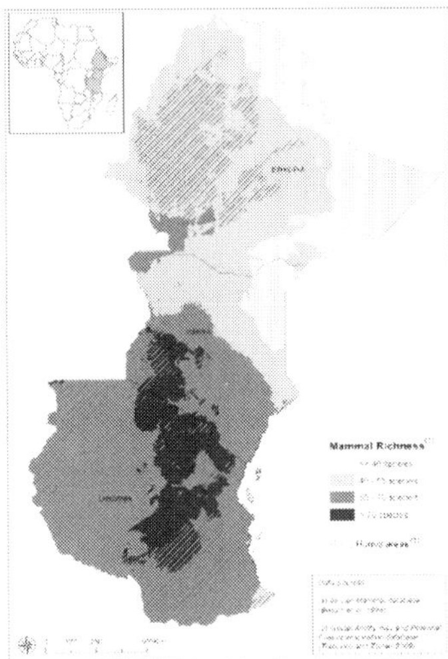

Figure 2: The mammal diversity in East Africa.

There has been some debate regarding the competition between livestock producers and biodiversity. Some researchers suggest that pastoral land use has created some of the most biologically diverse savannah ecosystems in the world (Fratkin 1997; Homewood and Rogers 1991;Little 1996) and that high intensities of livestock grazing do not adversely affect wildlife populations (Saberwal 1996). This is in contrast with the traditional conservation narrative which describes wildlife populations as threatened directly with extinction by human exploitation and indirectly by habitat degradation and fragmentation that resulted from increased human populations and their demands for development (Campbell 2002). Many people associate grazing animals with overgrazing and soil degradation. The Global Assessment of Soil Degradation (Oldeman et al. 1991), for example, estimates that 680 million ha of rangeland have become degraded since 1945, and (Dregne et al. 1991) argued that 73% of the world's 4.5 billion ha of rangeland is moderately or severely degraded.

However, savannah ecosystems are not the natural wildernesses that some conservationists believe (Little 1996). Over the 5,000 years since pastoralism emerged as a land use system in sub-Saharan Africa, pastoral natural resource management and herding strategies have modified ecosystems to such an extent that removal of pastoralism can be detrimental to grazing ungulates and rangeland diversity (Lamprey and Waller 1990) Pastoralist rangeland management practices that include the use of fire and periodic grazing to promote the growth of palatable grasses and reduce the encroachment of bushes have created conditions favourable for wildlife. In some locations, there is evidence of higher diversities and populations of wildlife in pastoral areas adjacent to national parks, than in the parks themselves (ILRI 2006).

There is an increasingly robust case for pursuing conservation objectives through pastoralism rather than seeing conservation and pastoralism as mutually exclusive pursuits. Positive relations between biodiversity and pastoral livelihoods have existed over the long term, and a picture of a harmonious co-existence has emerged (ILRI 2006).

Pastoralists have developed elaborate land-use strategies for conserving resources, and there are examples of grazing regimes, stocking regulations and pasture conservation within many pastoral societies (Ruttan and Borgerhoff Mulder 1999; McGahey et al. 2008). There is also increasing demonstration that conservation can play a significant role in sustainable development in the drylands. Indeed, pastoralists in industrialised countries increasingly rely on conservation-related incomes to bolster their livestock enterprise and to provide a diverse income portfolio in times of financial stress (WISP 2008). In some countries in Eastern Africa, the revenue from big-game tourism is comparable with the contribution of pastoralism. Kirkbride and Grahn (2008) report that tourism brings in annual returns of $900 million to $1.2 billion to Tanzania's economy and represents 13% of Kenya's GDP There are a number of examples of benefit sharing from national parks, wherein the park authorities effectively compensate neighbouring pastoralists for tolerating wildlife on their land, thereby securing dispersal areas outside the national parks and keeping wildlife corridors open (Horan et al. 2009; Rodríguez et al. 2012). There are also numerous examples of pastoralists initiating conservation-related businesses, by creating favourable conditions on their land for investment by tourism ventures (Manyara and Jones 2007; Groom and Harris 2008; Homewood et al. 2009). Payments to livestock herders for the ecosystem services generated through their land uses are currently being made in lands adjacent to Kenya's Maasai Mara National Reserve. The pastoral landowners leasing their land to the Olare Orok Conservancy, for example, currently receive a payment of US$43 ha^{-1} $year^{-1}$ (Bedelian in press).

Tourism is not the only way to benefit from biodiversity: conservation and efforts are underway to promote a range of other conservation-based incomes. These include marketing of natural products that are found in the drylands, such as gums, resins and medicinal plants. Environmental services can also be marketed, although there are few precedents to draw on yet from in the Eastern and Central African regions.

The Wider Development and Policy Context

To take full advantage of the potential biodiversity-pastoral synergies, it will be crucial to put in place supporting policies. They need to be embedded in the context of overall pastoral development and take the development of the whole dryland area environment into account.

A variety of agricultural production systems can be found in the drylands. The most widely spread livelihood strategy involves pastoral livestock production. Pastoral communities are however under pressure because of human population growth, breakdown of traditional management systems and exclusion from the wetter and more fertile parts of territories previously available for pastoral exploitation. These territories are encroached on by dryland agriculture or irrigation, which typically excludes grazing (Galvin 2009). In general, people in the Eastern African drylands on average lag far behind on human well-being, and many of its people live below the poverty line (UNDP 2008; WRI et al. 2007; Henninger et al. 2010). (Homewood et al. 2009) describe an immense variability of the poverty levels. Spatial variation exists between countries and locations. In addition, a small number of people with high incomes cause the mean income to be skewed upwards. This rather gloomy situation calls for development action. There is an urgent need for investment, particularly in education, health services, transport systems, local institutions, security and legal services.

In very broad terms, there are three different development pathways for the pastoral populations. One obvious option is to pursue development and enhance the welfare of the population through increasing the benefits and income generated from pastoral livestock production. Households can also opt to complement their income from livestock production through a diversity of alternative activities. Some herders remain in the sector but diversify their income while sustaining their pastoral livelihood. This is what we call a diversified pastoral livelihood. Finally, there is potential to migrate out of pastoralism into non- or marginally livestock-related activities. This can be pursued in the pastoral areas or through a complete outward migration out of pastoral areas. There is a variety of alternative land uses possible for both pastoral diversification and exit from pastoralism. They include - but are not limited to - crop agriculture, carbon sequestration, biofuel production, collection and sale of natural products and wildlife tourism. In reality, the dynamics and trends in pastoral livelihood strategies are much more complex and heterogeneous than this simple categorisation suggests, but the description of those goes far beyond the scope of this paper.

These differential development pathways often pose competing demands on the natural resources, especially in some of the key resource areas. Some of the wetlands, flood plains or other wetter sites form for example crucial grazing areas for the pastoralists but are also the first target for diversification in the form of crop agriculture. The multiple demands from different potential land uses have to be balanced carefully by planners and users, and the overall landscape taken into account if pastoralism as a livelihood has to survive and its synergies with biodiversity fully exploited.

Although the case for pastoralism can be made convincingly from an economic, environmental and social rights-based perspective, the prevailing policy environments are often detrimental to pastoralism and are, as a consequence, hampering the positive interaction of pastoralism with biodiversity. Policy makers often see conservation and agricultural production objectives as mutually exclusive and favour other land uses, some of which might be less environmentally sustainable and less economically viable in the drylands. To address this imbalance, it is necessary to identify sustainable investment options that re-enable pastoralists to invest in both livestock keeping and biodiversity conservation. Most governments in the region are struggling to get the right mix of policies that will address and spur accelerated pastoral development and protect biodiversity in the drylands.

The policy status in the East African region is rather differentiated. Countries like Kenya are coming up with strong policies to integrate drylands and pastoralism into the mainstream economy. This can be seen in the Vision 2030 which includes arid and semi-arid lands' unique needs and in the establishment of a Ministry of State for the Development of Northern Kenya and Other Arid Lands. Ethiopia and Tanzania, on the other hand, appreciate pastoralism in some policy documents and express contrary statements in official statements and other policy documents (MOAC 1997; SPILL and Mtatifikolo 2005; PASDEP 2006). It should be noted that there is an increasing appreciation and recognition of pastoralism at the regional level. In Africa, the African Union (AU) Policy Framework on Pastoralism and AU Framework and Guidelines for Land Policy in Africa are some of the continental policy instruments providing a generic framework for the countries to use to address the pastoralist issues. In East Africa, the East Africa Protocol on Environment and Natural Resources, the East Africa Climate Change Policy and the Intergovernmental Authority for Development (IGAD) Livestock Policy Initiative are some of the key instruments providing positive guidelines on pastoral issues. Though a mix of sometimes contradicting policies and laws can be found at the country level, the mere presence of positive policies is an improvement over the lack of or inappropriate policies that were common in the region in the past ten or so years. Research findings that challenge some of the critical negative perceptions on pastoralism, such as the 'tragedy of the commons', are finally trickling down and influencing the policy discourse

Taking a critical look at the economic policies for Kenya, Ethiopia and Tanzania, it can be seen that in terms of economic development planning, pastoralists have started to be recognised, and the need for special attention to dryland development is also captured. The ambiguity and mixed nature of policy and policy statements, however, indicate that not all key actors in the policy arena have been reached and that a lot more needs to be done to convince all policy makers that pastoralism is sustainable and economically viable. In addition, many policies do not have the strategies for implementation, and when they do, there is no guarantee of budgetary allocation nor a clear plan for implementation at different levels (local and national). This is not unexpected given that the policy-making process is, by no means, as rational as it is often portrayed to be. The outcomes occur as a result of complicated political, social and institutional processes which are at best described as evolutionary (Sutton 1999). Fortunately, the growing recognition that policy processes are complex, multidimensional and unpredictable (Omamo 2003; Young 2006) is encouraging mechanisms that promote the use of research-based evidence in development policy. This paper hopes to contribute to this process by arguing that pastoralism can be both ecologically and economically sustainable and providing a framework for evaluating investment and policy options in the drylands. Special attention is hereby given on how best to support biodiversity and pastoralism.

Spatial Heterogeneity

The drylands in Eastern Africa are highly heterogeneous. Climatic conditions for example range from hyper-arid to sub-humid (Trabucco and Zomer 2009) and vary considerably in rainfall variability (Jones and Thornton 2010). Soil characteristics and fertility show highly varied spatial patterns (Sanchez et al. 2003). A lot of variation can also be seen in terms of socio-economic variables. Population density in these drylands for example ranges from almost zero to more than 200 people per square kilometre (CIESIN et al. 2004). Travel times to settlements of 50,000 or more inhabitants are estimated to reach as long as 36 h (Nelson 2008). Rangeland landscapes and the communities inhabiting them are not all the same and will respond differently to both management practices and changes in the environment. As in any other rural development planning process, it is of crucial importance to take this complexity and heterogeneity into account when planning dryland development strategies. It influences the applicability and impacts of interventions, as well as the need for specific investments and policy support.

From an agricultural development perspective, absolute and comparative advantages of different communities in different locations are fundamentally important frames for designing development strategies (Chamberlin et al. 2006). The heterogeneity can be described in terms of many different variables, and the opportunities for development in dryland areas are determined by a multitude of factors, many of which may be closely associated. We focus in this paper on the gradients of agricultural production potential, market access and human population density because of their strong influence on the opportunities present and investments needed. Drylands are characterised by low and erratic precipitation that results in relatively low and unpredictable levels of crop and livestock production. Climatic conditions for example range from hyper-arid to sub-humid and vary considerably in rainfall variability. Pastoralists have developed unique mechanisms to cope with low and sporadic rainfall. They can maintain high populations of domestic herbivores sustainably if they have ensured and flexible access to the different habitats and resources in a given area (Niamir-Fuller 1999). The biomass production of pasture decreases, but its nutritive value increases with increasing aridity; pastoralists take advantage of this uneven distribution of plant productivity through livestock mobility, something that crop agriculture cannot achieve (Breman and de Wit 1983). The survival of herds depends on the pastoralists' willingness and capacity to move (Gallais 1977), and an extensive spatial scale of exploitation is a prerequisite for a successful pastoral system (Ellis and Swift 1988). The potential for crop agriculture on the other hand typically increases with humidity. The spread of crop production into drier lands can however hinder the mobility of pastoralists and also increase the conflicts between herders, farmers and wildlife. As crop cultivation moves into drier areas, it typically exploits key resource patches that are vital to pastoral production, removing a small but essential component from the bigger pastoral system. This creates a major externality that is seldom factored into the cost-benefit analysis of land use change (Davies and Hatfield 2008).

Proximity to markets and urban centres affects the number and range of options open to those interested in livelihood diversification. According to the research undertaken by Little (2005) in the Horn of Africa, pastoralists residing less than 40 km from towns typically have more alternative income-generating options than those living further away. However, according to (Little et al. 2008), there appears to be a trade-off between the disruptions to the pastoral production system brought about by restricted mobility and increased benefits of access to markets. They argue that the opportunities to move opportunistically in response to unpredictable rainfall patterns and forage production are more constraining near towns where markets are found but are more favourable in remote rangeland zones.

High population density forces pastoral people to diversify their livelihoods. This may be a diversification of pastoral livelihoods, including various natural resource or non-natural resource-based diversification options or exit out of the pastoral lands. Access to credit to enable investments would facilitate such diversification. Population density is also a proxy for the availability of labour, which is an important input in pastoral systems but might especially become a constraint when pastoralists diversify into non-land-related activities. Higher population density may enable labour-intensive livelihoods and land management approaches (Chamberlin et al. 2006) and may stimulate the development of local markets and infrastructure. It also increases the local demand and is likely to reduce transaction costs (Pender et al. 2006).

Policy Options for Supporting Pastoralism and Biodiversity

Policy has implications for the use and management of the rangelands. Policies can either support or constrain the interaction between pastoralism and biodiversity conservation. The policy environment in Eastern Africa has tended to be inappropriate for the proper functioning of the positive relationship between pastoralism and biodiversity conservation. Numerous policies exist but are often uncoordinated even within sectors and fall short on their evidence base (Chabari 2009; Manzano 2012). In addition, these policies are often not implemented (Hesse and Ochieng Odhiambo 2009).

Policies that support biodiversity investments will more accurately reflect the diverse livelihood portfolios of the region's pastoralists. This will be instrumental to preventing over-reliance on one or other economic activity and thereby will assist pastoralists to manage risk and to construct more resilient livelihoods. Policy support for a wider range of livelihood options will also help overcome the reliance on a narrow range of resources and thereby mitigate over-exploitation of those resources. In both cases, policy frameworks are only part of the solution, and it is recognised that further attention is also required to overarching socio-economic, political and institutional issues.

Depending on a location's position along the aridity, population density or market access gradient, supporting 'pastoralism and biodiversity' as a package, necessitates different investment and development actions and policy support.

Aridity

It is the mobility and flexibility of pastoral production systems that enable them to make the best use of the patchy and unpredictable environments that prevail in drylands (Gallais 1977; Ellis and Swift 1988; Behnke et al. 1993; Nori 2007; Krätli and Schareika 2010). In the landscape-scale management of pastoralism, cultivable or irrigable patches are only one component, compared to the total dependence of crop production on those patches and the associated opportunity cost of abandoning the remaining rangelands (Behnke and Kerven 2012; Niemi and Manyindo 2010). To ensure the resilience, integrity and sustainable management of rangeland ecosystems, these need to be managed at the ecosystem scale. Frequently, this does not happen, and rangelands become fragmented, disconnected and poorly managed. If the policy objective is to support 'pastoralism and biodiversity', pastoral mobility needs to be enabled, and integrated landscape-scale management strengthened. Table 1 below lists some of the interventions required in the more humid as well as arid environments.

Table 1: Investments and policies required along the aridity continuum

Dry sub-humid	Continuum	Arid or semi-arid
Protect access to communal resources (water and seasonal forage for wildlife and livestock) Ensure effective regulation of transhumance Compensation for wildlife damage Focus on diversification strategies to protect biodiversity Promote crop-livestock systems Promote crops that complement 'pastoralism and biodiversity'		Enable livestock mobility to maintain livestock-biodiversity corridors Strengthen communal resource management Focus on diversification to promote resilience Develop integrated pasture-water monitoring and management systems and institutions Provide infrastructure for integrated rangeland-water management Risk management and safety nets (e.g. insurance)

Notenbaert *et al.*

Notenbaert *et al. Pastoralism: Research, Policy and Practice* 2012 2:14, doi: 10.1186/2041-7136-2-14

It is assumed that crop cultivation in dry sub-humid areas is to some extent inevitable. Where crop cultivation is practised, close integration with livestock keeping could be promoted, through for example fodder production, ensuring access to water resources and seasonal forage and the regulation of transhumance. Further, the soils of a rangeland get easily exhausted and therefore must rely on fertiliser supplements to support continuous crop production (Okello and Grasty 2009). Supporting investments and policies need to be put in place to avoid abandonment of agricultural fields, and the consequent degradation that may take long to restore. To reduce the human-wildlife conflict, it might be necessary to compensate for wildlife damage.

At the drier end of the spectrum, the focus is on increasing resilience, through risk management, diversification of the pastoral livelihoods and holistic natural resource management. Species density may increase with humidity or may peak in the semi-arid zones, but species endemism may increase with aridity, whilst plant and animal species diversity is not so linear (Huston 1994) and species market value may increase with aridity. Examples of the latter include pharmaceutical properties as, e.g. diverse species in other regions such as North Africa (Hudge et al. 2002; Neffati et al. 2009), or touristic values as, e.g. in Samburu National Park in Kenya, where their 'special five'[a] - five dryland endemisms not found in other popular Kenyan parks - constitutes its major attraction. This higher market value with increased aridity is a clear opportunity for diversification of income associated with the drylands' biodiversity. In this context, the choice of appropriate technology and a strong emphasis on sound environmental management is especially vital to ensure sustainability, suitability and cost-effectiveness in the prevailing environment (UNDP 2009). This is to avoid situations as occurred in Kenya where *Acacia turkanensis* and *Acacia scabrifolia*are of urgent concern as a result of indiscriminate harvesting without any regard to conservation, or in another case, the deforestation in the habitats of gum and gum resin-producing tree species in Somalia (Wren 2009).

[a]The Samburu 'special five' - reticulated giraffe, long necked Gerenuk antelope, Grevy's zebra, Somali ostrich and Beisa oryx - deliberately contrasts with the popular 'big five' that is the flagship of Kenyan wildlife industry - lion, leopard, rhino, elephant and buffalo.

Population Density

Moving along the continuum from low to high population density, the emphasis on each of the investments or policies that we highlight in Table 2 should gradually increase or decrease.

Table 2: Investments and policies required along the population density continuum

Low population density	Contnuum	High population density
• Strengthening of the pastoral economy • Social services (including security) • Provision of infrastructure for tourism and market opening • Diversify to include biodiversity-related investments		• Strengthen land rights and NR governance against encroachment, fragmentation and resource pressures • Protect biodiversity • Rangeland rehabilitation • Diversify the economy through greater commercialisation and non- NR investments • Urban investments

Notenbaert *et al.*

Notenbaert *et al. Pastoralism: Research, Policy and Practice* 2012 **2**:14, doi:10.1186/2041-7136-2-14

As population density increases, greater emphasis is needed on diversifying the economy into non-natural resource-based activities. This will lessen the pressure on the natural resource base in general and on biodiversity in particular. The urban economy could be strengthened, so that a section of the population can successfully exit out of pastoralism, through providing access to credit and education complemented by infrastructural investments. As permanent settlements appear and continue to grow, there is a need to ensure pastoral mobility and access to key natural resources if pastoralism as a livelihood strategy is to survive. Strengthening the urban economy therefore needs to go hand in hand with regional planning so that rural development is not compromised. There is an urgent need to plan and guide the currently spontaneous and uncontrolled process of pastoral urbanisation (Little et al. 2008). High population density typically puts high pressure on biodiversity. The delineation and protection of conservation areas can contribute to the protection of biodiversity and ecological functioning, especially if connectivity is taken into account (Horan et al. 2009). As population density increases, typically less range land can be found. Development actions in these areas would still be to the benefit of biodiversity and pastoralism in the larger area if (1) actions for guaranteeing pastoral mobility and favourable land tenure and, consequently, habitat connectivity of ecosystems are undertaken; (2) advantage is taken from the increasing inflow of pastoralist population into the sprawling urban areas in order to promote commercialization of products from their areas of origin (Wanyoike et al. 2010). Commercialization strategies of products promoting biodiversity conservation will only work if financial services and certain investments in marketing facilities are guaranteed. Water provision to urban areas will need an integrated approach to guarantee ecosystem functioning and preservation of biodiversity and pastoralist livelihoods, especially in the more arid areas.

Market Access

Poor infrastructure, and insecurity, increases the costs and risks of livestock trading in remote areas (Barrett 2001). Smaller distances to the market are associated with higher probability of participation in the market, mainly due to high producer margins and low variable transaction costs (Ngeno et al. 2010). Similarly, greater proximity to markets increases the number and range of options open to those interested in livelihood diversification, leading to more possibilities for tapping in to biodiversity resources. Evidently, there are different market challenges to address for people living closer to markets. It is, for example, important to put policies and institutions in place that remove 'anti-competitive' bottlenecks, such as market exclusions and distortions by trader cartels (Barrett and Luseno 2004).

The transaction costs associated with distance from markets and the need to sell in bulk could be a disincentive to diversification: the more economic activities that are engaged in, the greater the cumulative transaction costs. Hence it may make sense to invest in specialist pastoral production.

Table 3 below summarises the investments and policies required if the policy objective is to support 'pastoralism and biodiversity' along the market access continuum.

Table 3: Investments and policies required along the distance to the market continuum

Far from markets	Continuum	Close to markets
Strengthen market chain connectivity Collective marketing Recognise opportunities for trans-boundary trade Infrastructure Market information Exploit the economy of scale in marketing Address asset liquidity and relate to banking services Invest in processing or transport of perishable products (especially milk)		Diversification of production (both in pastoralism and into non- pastoral activities) Private rather than collective market-enterprise Legislation to promote competition (e.g. avoid cartels)

Notenbaert *et al.*

Notenbaert *et al. Pastoralism: Research, Policy and Practice* 2012 **2**:14, doi:10.1186/2041-7136-2-14

DISCUSSION

An optimised use of the dryland areas to conserve biodiversity as well as encourage continuation of pastoralism in Eastern Africa necessitates careful regional land use planning, taking into consideration trade-offs at the landscape scale. The impacts of different livelihood options will have to be evaluated against different objectives, such as increasing food production, enhancing livelihoods (in terms of, for example, equity, health, income or food security), and maintaining biodiversity or environmental sustainability, and weighted accordingly.

Another challenge is presented by spatial and temporal scales. Interventions with positive outcomes at the local level often have disastrous effects when implemented at a larger geographical scale. Short-term benefits can be outlived by long-term negative consequences. Small-scale cropping for example offers the opportunity of a relatively stable income addition for pastoralists. Often it is, however, practised in key dry season grazing areas and around water points. In that case, it negatively affects the ability of herders to access key resources during periods of need, making pastoralism less sustainable over time, as well as generating local disputes and destructive conflicts over these key resources (Little 2009). Long-term negative outcomes include loss of biodiversity, soil compaction, the emergence of new pests and diseases, and nutrient mining (Watson and van Binsbergen 2006). The protection of wildlife through the delineation of national parks often excludes these areas from exploitation and restricts pastoral mobility, while only few of the benefits flow back to the pastoralist communities (Homewood et al. 2009; Kabiri 2010). In addition, the development of pastoral livelihoods typically involves a mix of interventions. The complementarities between different forms of land use practices therefore need to be taken into account. Policy and investment should support a diverse range of complementary land uses, as opposed to just one or the other land use option. It is possible that the aggregate values of multifunctional rangelands - pastoralism, biodiversity conservation, rangeland products and so on - can best be exploited by optimising their combined production, rather than maximising the production of one or the other at the cost of the remainder (Aboud et al. 2012).

The concept of heterogeneity can help planners and decision makers thinking through the types of investments and supporting policies needed, when evaluating the wide variety of livelihood options and land use systems. The differences along different axes favour certain investment options and define policy constraints that need to be addressed. Many more criteria could be evaluated and bring out other important factors. 'Pastoralism and biodiversity' remains an agro-ecosystem that requires system-based holistic planning. Independent of the level of aridity, population density and accessibility, there must be emphasis on landscape-scale land use planning. The complementarities between pastoralism and biodiversity are not guaranteed. They require special attention and investments and policies supporting these overlapping land uses. Examples from Kenya and Eastern Africa (Aboud et al. 2012; Flintan 2012) illustrate some of the ways forward: secure land rights, investment in local institutions, investment in ecotourism and investment in harvesting natural products such as henna or gum arabic. The future of the drylands will in the end largely depend on the development paradigms, we as a society, decide to follow. Will there be breakthroughs in payment schemes for ecosystem services in these areas (Reid et al. 2004)? The answers to these questions together with global trade and power relationships will have significant impacts on how some of these regions will develop and how much advantage will be taken from the potential complementarities between pastoralism and biodiversity.

Investments in infrastructure produce improved services for people and develop markets, but in the long term as investors buy large areas of land that can be put into alternative production, this could be at the expense of grazing land and biodiversity (Cotula et al. 2009; Oakland Institute 2011). If at the same time, payments for ecosystems services (PES) could be put in place for carbon, wildlife and the protection of water sources, pastoralists could perhaps manage livestock populations to optimise the whole ecosystem for the conservation of biodiversity. Care needs to be taken that these payment schemes do not lead to inequity as only the well-educated or more resourceful have the information to access payments for ecosystem services. Experience in Kenya, for example, shows that money generated by parks and community sanctuaries from tourism revenue mostly go to local elites, foreign tour investors or the government (Norton-Griffiths and Said 2010; Thomson and Homewood. 2002; Sindiga 1995). The participation of poor households in PES is limited by, among others, high transaction costs, institutional and technical barriers, lack of information, and weak capacity for negotiation (Iftikhar et al. 2007; Kosoy et al. 2005; Miranda et al. 2003; Zbinden and Lee 2005). This last issue again brings out the need to include communities in the planning. Pastoralism and pastoral lifestyles are unique and tailored to inhabit and use the drylands as efficient as possible, through use of traditional knowledge system and cultures. This resilient and adaptive knowledge of the pastoral people should be incorporated in the national policies and strategies where appropriate.

CONCLUSIONS

If the political agenda is to support the complementarities between pastoralism and biodiversity, there are a number of key investment and policy priorities. Ecologically and economically sustainable livestock production in the drylands requires public investment in transport and market infrastructure, credit facilities, health and education adapted to mobility in combination with policies supporting environmental protection. Drylands are, however, not homogeneous areas but contain diverse bio-physical and socio-economic realities. Land use planning and livelihood strategies, investments and policies need be matched to this diversity. Drylands are also highly complex systems operating at a different scale. Trade-offs at landscape need to be carefully examined when designing development strategies.

The case for pastoralism and biodiversity as a package has been made by others (e.g. Aboud et al. 1996; Neely et al. 2009; Norton-Griffiths and Said 2010), and this paper offers some options for realising this synergy. However, policies can endorse this complementarity, but a great deal more is required, particularly in terms of private investment, access to financial services, and training and education. More detailed research and experience-sharing are required to improve our understanding of the conditions under which the 'pastoralism and biodiversity' goals can be practically realised on the ground.

AUTHORS' CONTRIBUTIONS

AMN was responsible for the spatial analysis and took a lead in describing the study site, writing up the results, discussion and conclusion. JD was the lead writer of the sections on biodiversity and the positive relationship it exhibits with pastoralism. Together with PM, he listed the investments and policies required to support biodiversity and pastoralism. JDL and MS provided text for the background section and the section on the positive relationship between biodiversity and pastoralism, as did AA. SO reviewed the current policies, while MH and MW were instrumental in the initial shaping of the spatial framework In addition, MH contributed to the discussion section. All the authors have read and approved the final manuscript.

REFERENCES

1. Aboud, A, F Abdikadir, and M Hileman. 1996. Pastoralist life-styles and development paradigms: The contradictions and conflicts. Research document. Nairobi, Kenya: Kenya Pastoralist Forum.
2. Aboud, A, P Kisoyan, M Said, A Notenbaert, J de Leeuw, J Gitau, P Manzano, J Davies, G Roba, S Omondi, and M Odhiambo. 2012. Natural resource management and biodiversity conservation in the drylands of Eastern and Central Africa. Entebbe: Association for Strengthening Agricultural Research in Eastern and Central Africa (ASARECA).
3. Barrett, C. 2001. Livestock pricing and markets performance. Research brief 01-05-PARIMA. Global Livestock Collaborative Research Support Program (GL-CRSP). Davis: University of California.
4. Barrett, C, and WK Luseno. 2004. Decomposing producer price risk: a policy analysis tool with an application to northern Kenyan livestock markets. Food Policy 29: 393–405.
5. Bedelian, C. 2999. Conservation and tourism on privatised land in the Mara. Kenya: the case of conservancy land leases. Land Deal Politics Initiative Working Paper. University College London. in press
6. Behnke, R, and C Kerven. 2012. Counting the costs: replacing pastoralism with irrigated agriculture in the Awash valley. In Pastoralism and development in Africa: dynamic changes at the margin, ed. Catley A, Lind J, Scoones I, 57–70. London: Earthscan from Routledge.
7. Behnke, R, I Scoones, and C Kerven. 1993. Range ecology at disequilibrium: new models of natural variability and pastoral adaptation in African Savannas. London: Overseas Development Institute.
8. Boitani, L, F Corsi, A De Biase, I Carranza, M Ravagli, G Reggiani, I Sinibaldi, and P Trapanese. 1999. AMD African mammals databank. A databank for conservation and management of the African mammals. Report to the Directorate General for Development (GDVIII/A/1) of the European Commission. Roma: Instituto di Ecologia Applicata.10 CD-ROMs

9. Breman, H, and CT de Wit. 1983. Rangeland productivity and exploitation in the Sahel. Science 221: 1341–1347.
10. Briske, DD, JD Derner, JR Brown, SD Fuhlendorf, WR Teague, KM Havstad, RL Gillen, AJ Ash, and WD Willms. 2008. Rotational grazing on rangelands: reconciliation of perception and experimental evidence. Rangeland Ecology & Management 61: 3–17.
11. Bush, G, S Nampindo, C Aguti, and A Plumptre. 2004. Valuing Uganda's forests: A livelihoods and ecosystems approach. Kampala, Uganda: Technical report, European Union Forest Resources Management and Conservation Program.
12. Campbell, L. 2002. Conservation narratives in Costa Rica: conflict and co-existence. Development and Change 33: 29–56.
13. CIESIN (Center for International Earth Science Information Network) Columbia University CIAT (International Center for Research on Tropical Agriculture). 2004. Gridded population of the world (GPW), version 3 (beta). Palisades. NY: CIESIN, Columbia University.http://beta.sedac.ciesin.columbia.edu/gpw
14. Chabari, F. 2009. Food, livestock and pastoralism policies. Paper presented at the study conference on pastoralism in Eastern Africa and policy influencing in Europe. June 2009. Netherlands: The Hague.
15. Chamberlin, J, J Pender, and B Yu. 2006. Development domains for Ethiopia: capturing the geographical context of smallholder development options. Development Strategy and Governance Division, discussion paper 43. Washington DC: International Food Policy Research Institute.
16. Cotula, L, S Vermeulen, R Leonard, and J Keeley. 2009. Land grab or development opportunity? Agricultural investment and international land deals in Africa. London/Rome: International Institute for Environment and Development/Food and Agriculture Organisation/International Fund for Agricultural Development.
17. Davies, J, and R Hatfield. 2008. The economics of mobile pastoralism: a global summary. Nomadic Peoples 11: 91–116.
18. Davies, J, M Niamir-Fuller, K Kerven, and K Bauer. 2010. Extensive livestock production in transition: the future of sustainable pastoralism. In Livestock in a changing landscape, Volume 1,

Drivers, consequences, and responses, ed. Steinfeld H, Mooney HA, Schneider F, Neville LE, 285–308. Washington, DC: Island Press.

19. Dregne, H. 1991. Arid land degradation - a result of mismanagement. Geotimes 36 (6): 19–21.
20. Ellis, J, and D Swift. 1988. Stability of African pastoral ecosystems: alternate paradigms and implications for development. Journal of Range Management 41 (6): 450–458.
21. FAO. 2009. Sustaining communities, livestock and wildlife. A decision support tool. Food and Agriculture Organization: Rome.
22. Flintan, F. 2008. Women's empowerment in pastoral societies. Nairobi: International Union for Conservation of Nature-WISP.
23. Flintan, F. 2012. Making rangelands secure: past experience and future options. Rome: International Land Commission.
24. Fratkin, E. 1997. Pastoralism: Governance and development issues. Annual Review of Anthropology 26: 235–261.
25. Gallais, J. 1977. Stratégies pastorales et agricoles des Sahéliensiens durant la Sécheresse 1969-1974. Travaux et Documents de Géographies Tropicales no. 30. Paris: Centre d'Etudes de Geographie Tropical (CEGET).
26. Galvin, K. 2009. Transitions: pastoralists living with change. Annual Review of Anthropology 38: 185–198.
27. Groom, R, and S Harris. 2008. Conservation on community lands: the importance of equitable revenue sharing. Environmental Conservation 35: 242–251.
28. Henninger, NF Landsberg with the Ministry of Agriculture Animal Industry and Fisheries, Uganda Uganda Bureau of Statistics Food and Agriculture Organization of the United Nations International Livestock Research Institute. 2010. Mapping a better future: spatial analysis and pro-poor livestock strategies in Uganda. Washington DC: World Resources Institute.
29. Hesse, C, and M Ochieng Odhiambo. 2009. Strengthening pastoralists' voice in shaping policies for sustainable poverty reduction in ASAL regions of East Africa. Presented at the study conference on Pastoralism in Eastern Africa and Policy Influencing in Europe. June 2009. Th Hague, the Netherlands.

30. Homewood, K, and W Rogers. 1991. Maasailand ecology: pastoralist development and wildlife conservation in Ngorongoro, Tanzania. Cambridge: Cambridge University Press.
31. Homewood, K, P Kristjanson, and PC Trench. 2009. Staying Maasai? Livelihoods, conservation and development in East African rangelands. New York: Springer.
32. Horan, RD, JF Shogren, BM Gramig, et al.. 2009. Conservation payments to reduce wildlife habitat fragmentation and disease risks. In Payment for environmental services in agricultural landscapes, ed. Lipper L, 103–132. Natural Resource Management and Policy.
33. Hudge, VS, MR Shobhane, MV Dhoble, and AM Dhopte. 2002. Suitable medicinal and aromatic crop plants for dryland farming. In Agrotechnology for dryland farming, ed. Dhopte AM, 593–612. Jodhpur, India: Scientific Publishers.
34. Huston, MA. 1994. Biological diversity: the coexistence of species on changing landscapes. Cambridge: Cambridge University Press.
35. Iftikhar, UA, M Kallesoe, A Duraiappah, G Sriskanthan, SV Poats, and B Swallow. 2007. Exploring the inter-linkages among and between compensation and rewards for ecosystem services (CRES) and human well-being: CES Scoping Study Issue Paper no. 1. ICRAF Working Paper no. 36. Nairobi: World Agroforestry Centre.
36. ILRI. 2006. Pastoralist and poverty reduction in East Africa. Conference, June. Nairobi Kenya: International Livestock Research Institute (ILRI).
37. IUCN (World Conservation Union) and UNEP-WCMC (United Nations Environment Programme's World Conservation Monitoring Centre). 2010. The World Database on Protected Areas (WDPA). Cambridge, UK: UNEP-WCMC. Available at: www.protectedplanet.net
38. Jones, P, and P Thornton. 2010. Global length-of-growing-period surfaces for a suite of Fourth Assessment climate models. Nairobi, Kenya: Digital data set, ILRI.
39. Kabiri, N. 2010. The political economy of wildlife conservation and decline in Kenya. The Journal of Environment and Development 19: 424–445.

40. Kirkbride, M, and R Grahn. 2008. Survival of the fittest: pastoralism and climate change in East Africa. Oxfam Briefing Paper 116, August. Oxford: Oxfam International.http://www.oxfam.org

41. Kosoy, N, M Martínez-Tuna, R Muradian, and J Martínez-Alier. 2005. Payments for environmental services in watersheds: insights from a comparative study of two cases in Central America, seminar on environmental services and financing for the protection and sustainable use of ecosystems. Geneva: Swiss Agency for the Environment, Forest and Landscape Geneva.10–11 October

42. Krätli, S, and N Schareika. 2010. Living off uncertainty: the intelligent animal production of dryland pastoralists. European Journal of Development Research 22: 605–622.

43. Lamprey, H, and R Waller. 1990. Thc Loita-Mara region in historical times: patterns of subsistence, settlement and ecological change. In Early pastoralists of south-western Kenya. Memoirs of the British Institute, ed. Robertshaw P, 16–35. Nairobi: British Institute in Eastern Africa.

44. Little, P. 1996. Pastoralism, biodiversity and the shaping of savanna landscapes in east Africa. Africa 66: 37–51.

45. Little, P. 2005. Unofficial trade when states are weak: the case of cross-border commerce in the Horn of Africa. Research Paper no. 2005/13. Helsinki, Finland: World Institute for Development Economics Research, United Nations University.

46. Little, PD. 2009. Hidden value on the hoof: cross-border livestock trade in Eastern Africa. 2009. Common Market for Eastern and Southern Africa Comprehensive African Agriculture Development Program, Policy Brief Number 2.

47. Little, P, J McPeak, C Barrett, and P Kristjanson. 2008. Challenging orthodoxies: understanding poverty in pastoral areas of East Africa. Development and Change 39 (4): 587–611.

48. Manyara, G, and E Jones. 2007. Community-based tourism enterprises development in Kenya: an exploration of their potential as avenues of poverty reduction. Journal of Sustainable Tourism 15: 628–644.

49. Manzano, P. 2012. Shaping policies: science-policy interface in natural resources management. In Conservation and sustainable development: linking practice and policy in Eastern Africa, ed.

Davies J, 107–126. New York and London: Taylor & Francis IDRC, Ottawa.

50. McGahey, D, J Davies, and E Barrow. 2008. Pastoralism as conservation in the Horn of Africa: effective policies for conservation outcomes in the drylands of Eastern Africa. Annals of Arid Zones 46: 353–377.
51. Miranda, M, I Porras, and M Moreno. 2003. The social impacts of payments for environmental services in Costa Rica. A quantitative field survey and analysis of the Virilla Watershed. Markets for Environmental Services. London: International Institute for Environment and Development.
52. MOAC. 1997. Agricultural and Livestock Policy. Ministry of Agriculture and Cooperatives. Dar-es-Salaam: United Republic of Tanzania.
53. Mortimore, M. 2005. Why invest in drylands? A report for the global mechanism of the United Nations Convention to Combat Desertification. Rome: UNCCD.
54. Nassef, M, S Anderson, and C Hesse. 2009. Pastoralism and climate change, enabling adaptive capacity. London: Regional Pastoral Livelihoods Advocacy Project, Nairobi and Overseas Development Institute.http://www.iied.org/pubs/pdfs/G02497.pdf
55. 2009. Review of evidence on dryland pastoral systems and climate change: implications and opportunities for mitigation and adaptation. Land and water discussion paper 8,Neely C, Bunning S, Wilkes A Rome: Food and Agriculture Organization of the United Nations.
56. Neffati, M, A Ouled Belgacem, and M El Mourid. 2009. The medicinal and aromatic plants sector in the drylands: a promising alternative for sustainable development and combating desertification in Tunisia. In The future of drylands, ed. Lee C, Schaaf T, 171–182. UNESCO/Springer: International Scientific Conference on Desertification and Drylands Research Tunis, Tunisia, 19-21 June 2006, Paris/Dordecht.
57. Nelson, A. 2008. Travel time to major cities: A global map of Accessibility. Ispra Italy: Global Environment Monitoring Unit - Joint Research Centre of the European Commission.http://www-tem.jrc.it/accessibility

58. Ngeno, V, B Langat, M Korir, E Ngeno, and M Kipsat. 2010. Probability of pastoralists participation in livestock markets in East Pokot district. Kenya: Proceedings of the 12th KARI biennial scientific conference, Kenya.8-12 November 2010

59. Niamir-Fuller, M. 1999. Managing mobility in African rangelands: The legitimization of transhumance. London: Intermediate Technology Publications.

60. Niemi, E, and J Manyindo. 2010. Economic importance of goods and services derived from dryland ecosystems in the IGAD region: case studies. Nairobi: IUCN.http://cmsdata.iucn.org/downloads/4615_fnl_case_studies_10_0421__6_.pdf

61. Nori, M. 2007. Mobile livelihoods, patchy resources and shifting rights: approaching pastoral territories. Rome: International Land Coalition.

62. Norton-Griffiths, M, and M Said. 2010. The future for wildlife on Kenya's rangelands: an economic perspective IN. In Wild rangelands: conserving wildlife while maintaining livestock in semi-arid ecosystems, ed. Toit J, Kock R, Deutch J, 367–392. New York: Wiley Interscience.

63. Oakland Institute. 2011. Understanding land investment deals in Africa: Ethiopia country report.http://media.oaklandinstitute.org/sites/oaklandinstitute.org/files/OI_Ethiopa_Land_Investment_report.pdf

64. Okello, M, and K Grasty. 2009. Contraction and status of Maasai lands as wildlife dispersal areas and implications for wildlife conservation in Amboseli ecosystem, Kenya. In Wildlife: destruction, conservation and biodiversity, ed. Harris JD, Brown PL, 49–96. Nova Science Publishers: A Wildlife Protection, Destruction and Extinction Nova Publishers Book Series.

65. Oldeman, , LR Hakkeling, and W Sombroek. Word Map of of the status of human-induced soil degradation, and explanatory notes. Wageningen, The Netherlands: Global Assessment of Soil Degradation (GLASOD), ISRIC.

66. Omamo, SW. 2003. Policy research on African agriculture: trends, gaps, and challenges, International Service for National Agricultural Research (ISNAR) Research Report No 21. The Hague: International Service for National Agricultural Research.

67. PASDEP. 2006. PASDEP - a plan for accelerated and sustained development to end poverty. Addis Ababa, Ethiopia: Ministry of Works and Urban Development.

68. Pender, J, S Ehui, and F Place. 2006. Conceptual framework and hypotheses. In Strategies for sustainable land management in the East African highlands, ed. Pender J, Place F, Ehui S, 31–58. Washington, DC: International Food Policy Research Institute.

69. Rodríguez, LC, D Henson, M Herrero, D Nkedianye, and R Reid. 2012. Private farmers' compensation and viability of protected areas: the case of Nairobi National Park and Kitengela dispersal corridor. International Journal of Sustainable Development and World Ecology 19: 34–43.

70. Ruttan, L, and M Borgerhoff Mulder. 1999. Are East African pastoralists truly conservationists? Current Anthropology 40 (5): 621–652.

71. Saberwal, V. 1996. Pastoral politics: Gaddi grazing, degradation and biodiversity conservation in Himachal Pradesh, India. Conservation Biology 10: 741–749.

72. Sanchez, P, C Palm, and S Buol. 2003. Fertility capability soil classification: a tool to help assess soil quality in the tropics. Geoderma 114: 157–185.

73. Savory, A. 1999. Holistic management: a new framework for decision making. Washington DC: Island Press.

74. Scoones, I. 1993. Why are there so many animals? Cattle population dynamics in the communal areas of Zimbabwe. In Range ecology at disequilibrium, ed. Behnke R, Scoones I, Kerven C, 62–76. London: Overseas Development Institute.

75. Scoones, I. 1995. Living with uncertainty: new directions for pastoral development in Africa. London: Intermediate Technology Press.

76. Secretariat of the Convention on Biological Diversity. 2005. Handbook of the convention on biological diversity. Montreal.

77. Sindiga, I. 1995. Wildlife-based tourism in Kenya: land use conflicts and government compensation polices over protected areas. J Tour Stud 1995 6 (2): 45–55.

78. Mtatifikolo, PSPILL (Strategic Plan for the Implementation of the Land Laws). 2005. Main consultancy report. Dar es Salaam:

Ministry of Lands and Human Settlements Development (MLHSD).

79. Stebbins, G. 1987. Coevolution of grasses and herbivores. Annals of the Missouri Botanical Garden 86: 75–86.

80. Sutton, R. 1999. The policy process: an overview, ODI working paper 118. London: Overseas Development Institute.

81. Swallow, B. 1994. The role of mobility within risk management strategies of pastoralists and agropastoralist. Gatekeeper series, SA47. London; International Institute for Environment and Development.

82. Thomson, M, and K Homewood. 2002. Entrepreneurs, elites and exclusion in Maasailand: trends in wildlife conservation and pastoralist development. Human Ecology 30. 107–138.

83. Trabucco, A, and R Zomer. 2009. Global aridity index (global-aridity) and global potential evapo-transpiration (global-PET) geospatial database. CGIAR Consortium for Spatial Information. http://www.csi.cgiar.org

84. UNDP. 2006. Making markets work for the poor.http://www.undp.org/drylands/docs/marketaccess/Making_Market_ Work_for_ Poor.pdf

85. UNDP. 2008. Human development report 2007/2008.

86. UNDP. 2009. Improving market access for dryland commodities in East Africa: synthesis report. http://web.undp.org/drylands/marketaccess-activities.html

87. UNEP (United Nations Environment Programme). 2007. Eastern Africa and biodiversity. In Encyclopedia of earth, ed. Cutler J, Washington, DC: Environmental Information Coalition, National Council for Science and the Environment.http://www.eoearth.org/article/Eastern_Africa_and_biodiversity. Accessed 5 December 2010

88. Vetter, S. 2005. Rangelands at equilibrium and non-equilibrium: recent developments in the debate. Journal of Arid Environments 62 (2005): 321–341.

89. Wanyoike, F, S Kaitibie, S Kuria, A Bruntse, IN Thendiu, DM Mwangi, and A Omore. 2010. Consumer preferences and willingness to pay for improved quality and safety: the case of fresh camel milk and dried camel meat in Kenya. In Demand

for livestock products in developing countries with a focus on quality and safety attributes: evidence from Asia and Africa. ILRI research report 24, ed. Jabbar MA, Baker D, Fadiga ML, 93–102. Nairobi: ILRI.

90. Watson, D, and J van Binsbergen. 2006. Life beyond pastoralism: livelihood diversification opportunities for pastoralists in Turkana district, Kenya. ILRI brief 3. Nairobi, Kenya: ILRI.
91. WISP. 2008. Policies that work for pastoral environments. A six country review of positive policy impacts on pastoral environments. Nairobi: The World Initiative for Sustainable Pastoralism, IUCN-ESARO.
92. Wren, S. 2009. Summary report on the potential for plant based rangeland enterprises in the drylands of Ethiopia. A study conducted for the ELSE consortium.http://www.elmt-relpa.org/FCKeditor/UserFiles/File/elmt/201002/summary%20report%20on%20the%20potential%20for%20planrt%20based%20rangeland%20enterprise%20in%20the%20drylands%20of%20Ethiopia.pdf
93. WRI (World Resources Institute) Department of Resource Surveys and Remote Sensing Ministry of Environment and Natural Resources, Kenya Central Bureau of Statistics Ministry of Planning and National Development, Kenya International Livestock Research Institute. 2007. Nature's benefits in Kenya: an atlas of ecosystems and human wellbeing. Washington DC: World Resources Institute.
94. Young, J. 2006. Bridging research and policy: the RAPID approach. Paper presented at the International conference African economic research institutions and policy development: Opportunities and challenges. Dakar, January 28-29 2005: Institutional Support for Economic Research in Africa (SISERA).
95. Zbinden, S, and D Lee. 2005. Paying for environmental services: an analysis of participation in Costa Rica's PSA program. World Development 33 (2 special issue): 255–257.

Chapter 8

Inter- and Trans Disciplinary Perspective on the Integration of Ecological Processes into Ecosystem Services Analysis in a Mountain Region

Robert Huber[1], Simon Briner[2], Harald Bugmann[3], Ché Elkin[3], Christian Hirschi[4], Roman Seidl[5], Rebecca Snell[3], and Andreas Rigling[1]

[1]Forest Dynamics, Swiss Federal Research Institute WSL, Zürcherstrasse 111, CH-8903 Birmensdorf, Switzerland

[2]Department of Environmental Systems Science, Agri-food and Agri-environmental Economics Group, ETH Zurich, CH-8092 Zurich, Switzerland

[3]Department of Environmental Systems Science, Forest Ecology, ETH Zurich, CH-8092 Zurich, Switzerland

[4]Department of Environmental Systems Science, Environmental Policy and Economics, ETH Zurich, CH-8092 Zurich, Switzerland

[5]Department of Environmental Systems Science, Natural and Social Science Interface, ETH Zurich, CH-8092 Zurich, Switzerland

ABSTRACT

Introduction

Ecosystem goods and services (EGS) studies have had little impact on policy processes and real-world decision-making due to limited understanding of the interactions and feedbacks among ecological, social and economic processes. Here we present an inter- and transdisciplinary analysis of global change impacts on EGS provision in a European mountain region. Our aim is to evaluate the projected influence of ecological, economic and social drivers on future EGS provision and to show possible ways to address the predominant limitations of EGS studies.

Methods

The integrated findings from ecological experiments, mechanistic models of landscape dynamics, socio-economic land-use models, policy analysis and transdisciplinary stakeholder interactions are presented consecutively. Four regionally downscaled global change scenarios, for a case study region near Visp, Switzerland (350 km^2), were used to examine the impacts of climate and socio-economic changes on four ecosystem services, i.e., food provision, timber production, net greenhouse gas emissions and protection from natural hazards.

Results

Our simulation results reveal four key aspects that influence the future provision of mountain EGS. First, we show the high spatial and temporal heterogeneity of EGS provision even in a small case study region. Second, we find that climate change impacts are much more pronounced for forest EGS, while changes to agricultural EGS result primarily from shifts in economic conditions. Third, our modeling results reveal the complex trade-offs associated with the different scenarios. Fourth, simulations illustrate the importance of interactions between environmental shifts and economic decisions. We discuss our simulation results with respect to both existing policy networks and trans disciplinary stakeholder interactions.

Conclusion

We describe a framework based on experiments and observations that effectively supports the integration of ecological processes into an integrative modeling chain of EGS provision in mountain regions, the political decision-making process and also trans disciplinary stakeholder interactions.

INTRODUCTION

The concept of ecosystem goods and services (EGS) provides a framework to analyze relationships between society and ecology in complex social-ecological systems (Reyers et al. 2010; MEA 2005). While there is no universally accepted typology of EGS (Haines-Young and Potschin 2009; Braat and de Groot 2012; Gómez-Baggethun et al. 2010), most EGS studies combine an evaluation of biophysical structures and processes with an evaluation of human, social and economic components that determine EGS benefits or values (Braat and de Groot 2012). Although a wide range of EGS studies have used this approach (reviewed in Seppelt et al. 2011), a challenge still remains, namely the effective integration of ecological, economic and social components into an internally consistent framework that can be used to project future changes in EGS and their response to different driving forces. For many stakeholders, including government agencies (Euliss et al. 2011), it is important to have robust and consistent approaches when assessing the effect of management actions on ecological processes and changes in EGS. Such approaches, however, are confronted with four main challenges. First, the ecological understanding of EGS is often limited (Kremen 2005), and there is a lack of quantitative relationships among biodiversity, ecosystem components and processes and services (de Groot et al. 2010). In addition, even if a mechanistic understanding of these processes exists, feedbacks between and within social and ecological systems are often ignored (Nicholson et al. 2009) or prone to inconsistencies (Cumming et al. 2005). Second, EGS studies often neglect the economic aspects of marginality, ecosystem transitions and substitution effects (Fisher et al. 2008; Bateman et al. 2011; Farley 2012), and sometimes fail to consider socially relevant and stakeholder-oriented approaches (Cowling et al. 2008; Reed et al. 2009). Third, valuation and monetization of EGS must be placed in a relevant socio-cultural context (Braat and de Groot 2012; Nelson et al. 2009) to ensure that the economic valuation of EGS is accurate and reflects regional characteristics (de Groot et al. 2012; Hein et al. 2006; Schaeffer 2008; Polasky and Segerson 2009). Fourth, while the EGS viewpoint facilitates an analysis of ecological processes from the ecological systems perspective, various scale differences and cross-scale dynamics (Young 2002; Cash et al. 2006) still make it difficult to obtain an integrative overview of ecological and human systems. Thus, due to a limited understanding of the interactions and feedbacks among ecological, social, economic and political processes, EGS studies have, historically, had little impact on policy processes and real decision-making (Daily et al. 2009; Carpenter et al. 2009). Consequently, a wide range of inter- and transdisciplinary frameworks have emerged in recent years, not only to analyze EGS in the context of social-ecological systems (e.g., Chapin et al. 2006; Wainger and Mazzotta 2011; Diaz et al. 2011; Oteros-Rozas et al. 2012; Polasky and Segerson 2009; Collins et al. 2011; Rounsevell et al. 2010; Kremen et al. 2007; Seidl et al. 2013), but also to meet the challenge of defining indicators that allow an understanding of the feedbacks and interactions in EGS analyses (van Oudenhoven et al. 2012; Müller and Burkhard 2012; Wainger and Mazzotta 2011). In addition, modeling studies that link ecological process models with socio-economic land-use models and incorporate feedback mechanisms (Claessens et al. 2009; Bithell and Brasington 2009; Schreinemachers and Berger 2011; Gibon et al. 2010; Huber et al. 2013a) have become state-of-the-art in ecosystem services analysis. This inter- and transdisciplinary study is part of the "Mountland" project, where we address global change impacts on EGS in three different European mountain regions (Huber et al. 2013b). The aim of the project was to identify land-use practices and alternative policy solutions that ensure the provision of EGS from agricultural land use and forestry in mountain regions subject to climate and land-use changes. The main results from this project, which was conducted in the three different case study regions, are summarized and synthesized in Huber et al. (2013c). The simulation results presented here complement this work by providing an overview of our integrative assessment considering all four scenario families in the Special Report on Emission Scenarios (SRES) from the Intergovernmental Panel on Climate Change (IPCC 2000). The value of this contribution lies in the illustration of a successful application of inter- and transdisciplinary research from joint scenario development to simulation results and policy analysis in one case study region. We analyze and describe the spatial and temporal variance and similarities between regionally downscaled SRES and the four associated climate scenarios with regard to the provision of four mountain EGS (i.e., food provision, timber production, net greenhouse gas emissions and protection from natural hazards) in the case study region Visp, Switzerland. The comprehensive nature of our analysis allows us to assess how and where within the landscape climate and socio-economic factors influence EGS provision. The integration of our interdisciplinary results as well as transdisciplinary stakeholder interactions permit us to illustrate the importance of considering ecological processes throughout an integrative assessment of EGS provision. The discussion illustrates how we used inter- and transdisciplinary research to overcome limitations and to further our understanding of the interlinked ecological, social and economic processes in EGS studies.

DATA, METHODS AND INTERDISCIPLINARY WORKFLOW

Case Study Region

Central Valais is a drought-sensitive, continental, inner-Alpine mountain area (Figure 1). It is suitable for studying the dynamics and diversity of mountain protection forests and grasslands and the influence of disturbances, in particular fire, insect and pathogen damage brought about by climate change (Rigling et al. 2013). The study area covers 350 km2 and has 15,000 inhabitants. Visp is a booming urban hotspot with industry and international traffic corridors. Saas Fee and Visperterminen are important destinations for summer and winter tourism. The Baltschieder valley is a remote, uninhabited Alpine side-valley with no infrastructure. Unproductive land accounts for 62% of the area, while 20% of the area is covered by forest and about 16% is used for agriculture. More than 98% of this agricultural land is grassland, and agricultural activities focus mainly on milk and meat production. Winter wheat and corn are the main crops cultivated in the bottom of the valley around Visp. On average, individual farmers cultivate less than 10 ha of agricultural land and keep around seven livestock units. Agricultural land use and forestry play an important role in the provision of recreation areas and habitats for plants and wildlife. Forest management is mainly in the hands of regional forest managers, with the primary aim of maintaining healthy forests that provide protection from rock fall and avalanche hazards. Timber production is of secondary importance due to high labor cost and low timber prices.

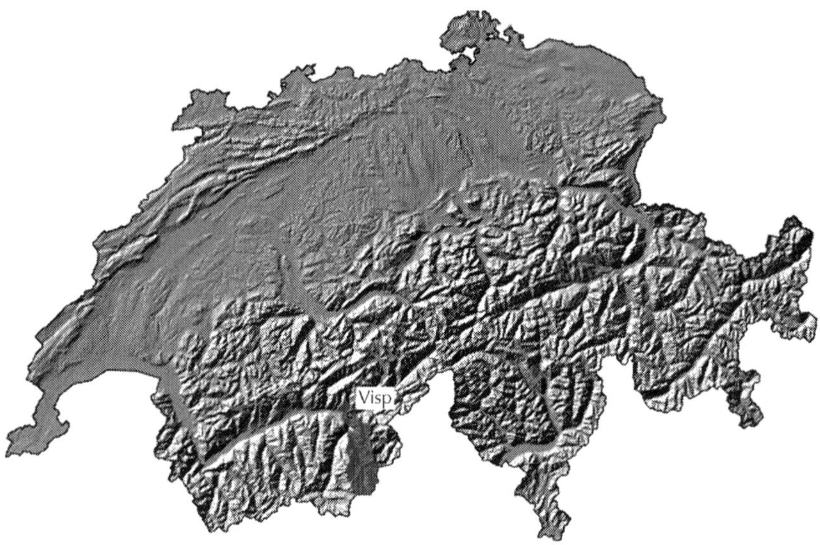

Figure 1: Map of Switzerland and the case study area Visp (in blue) in the Swiss Alps (provided by A. Psomoas, Swiss Federal Research Institute WSL).

Methodologies

The combination of different methodological approaches within the project allows us to produce a robust evaluation of future EGS provision in mountain regions under global change, taking interactions between the ecological, socio-economic and policy domains into account (Huber et al. 2013b). The main research findings from the Mount land project have been published in different interdisciplinary and disciplinary manuscripts (see Table 1). Thus, the workflow presented here does not describe every method in detail but instead focuses on the conceptual and practical link between the methodologies and the description of the interacting simulation frameworks. Figure 2 shows the general framework and the iterative procedure followed by the different research groups. More information on the technical integration of models and the corresponding assumptions can be found in the corresponding articles cited in Table 1.

Table 1: Research methodologies applied in the Mount land project

Research discipline and focus	**Methodology**	**Description**	**References**
Ecology (impact of global change on forest processes and ecosystem goods and services)	Experiments	Afforestation (irrigated/ non-irrigated, altitudinal gradients, transplantations)	Eilmann et al. 2010, 2011
		Rain-out shelter (under shelter/open land)	Eilmann and Rigling 2012
		Thinnings (understory/ overstory removal)	Rigling et al. 2013
	Monitoring	National forest inventory and plots from the long-term ecosystem monitoring program	Weber et al. 2008; Wermelinger et al. 2008; Zweifel et al. 2009; Rigling et al. 2010; Heiniger et al. 2011
		Additional temporal monitoring of plots (e.g., for insects, mistletoe, pathogens)	
	Modeling	Forest dynamics and future forest states were simulated using the landscape-scale model LandClim	Elkin and Bugmann 2010; Elkin et al. 2012, 2013

Socio-economic (land-use change analysis under global change)	Modeling	Modeling concomitant effects of climate and land-use change on the provision of ecosystem goods and services using an activity-based, spatially explicit economic land allocation model	Briner et al. 2012, 2013ab
	Formative scenario analysis	Constructing consistent regional multi-scale scenarios by transdisciplinary processes	Brand et al. 2013
Policy analysis (identification of best accepted policy strategies)	Policy network and policy preferences analysis	Network structures and governance modes in adaptation policies in the Canton of Valais; role of network governance in enhancing sustainable development in mountain areas	Ingold et al. 2010; Hirschi2010
		Network dynamics and policy preferences in Swiss agricultural and forest policy	Hirschi et al. 2013

Huber *et al.*

Huber *et al.* *Ecological Processes* 2014 3:9, doi:10.1186/2192-1709-3-9

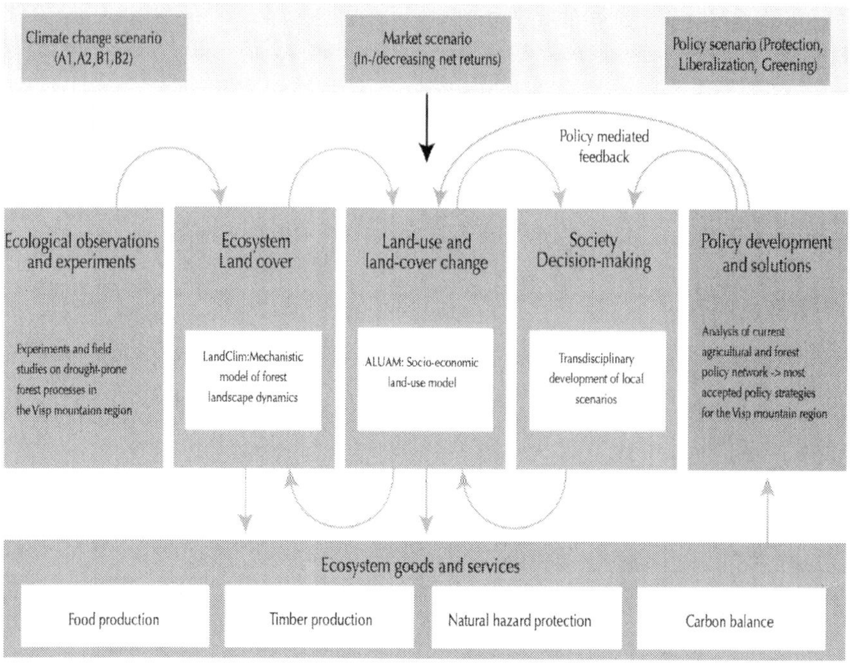

Figure 2: Conceptual framework of Mount land applied to the case study region Visp. Green fields refer to ecological research, blue to socio-economic research and orange to policy assessments. Source: Adapted from Huber et al. (2013b).

Scenario Development

We began our inter- and transdisciplinary research by formulating "context scenarios." In these scenarios, the consequences of global change at climate, market and policy levels are downscaled to the Visp region and summarized in regionally specific storylines consistent with the four IPCC SRES scenarios (Walz et al. 2013). We employed a formal technique of scenario construction which combines expert judgment with a quantitative, indicator-based selection algorithm. The resulting scenarios reflect the potential development pathways and interactions of major drivers for ecosystem service management and provide the input for a model-based analysis of ecosystem development and land-use decision-making.

Ecological Experiments and Modeling

To integrate knowledge of ecological processes into the Mountland model chain (green fields in Figure 2), findings from local ecological field studies and drought experiments (Brunner et al. 2009; Dobbertin et al. 2010; Eilmann and Rigling 2012; Eilmann et al. 2009, 2013; Rigling et al. 2013) were analyzed to improve the existing mechanistic forest model LandClim (Schumacher et al. 2004). LandClim is a spatially explicit process-based model that incorporates competition-driven forest dynamics and landscape-level disturbances to simulate forest dynamics at a landscape scale. LandClim was designed to examine the impact of climate change and forest management on forest development and structure (Schumacher and Bugmann 2006). The model simulates forest growth in 25 × 25 m cells using simplified versions of tree recruitment, growth and competition processes. Individual cells are linked by the spatially explicit processes of seed dispersal, landscape disturbances and forest management. Forest growth is determined by climatic parameters, soil properties, topography, and large-scale disturbances, as well as forest management. We used our findings from the ecological field studies and drought experiments in the Visp region to calibrate and validate the LandClim simulations. Thus, empirical information on the effects of climate change on drought-prone forests in the Visp region (e.g., germination, growth and mortality of different tree species) was integrated into LandClim. This improved the ability of LandClim to simulate forest dynamics and project the impact of climate and land-use change on the spatially explicit provision of EGS at the landscape scale (Elkin et al. 2012).

Socio-Economic Modeling

The simulation results from LandClim were used as input for the analysis of socio-economic impacts (blue fields in Figure 2). The economic land-use optimization model ALUAM (Briner et al. 2012) simulates the competition between different agricultural land uses and forestry. The model results explicitly consider feedbacks and trade-offs between the economic gains from primary production and the provision of EGS under different scenarios (Briner et al. 2013a, b). The simple, modular structure of ALUAM facilitates a linkup of economic aspects with ecological and biophysical elements and the integration of datasets from different sources. Since the model also permits the introduction of possible new activities (e.g., irrigation), it is suitable for the simulation of strategies for adaptation to global change. In ALUAM, we assume that economic agents are profit maximizers. Combined with limited resources, represented by model restrictions, this normative model approach incorporates the fundamental economic problem at the margins, which is to make the most of limited resources (Buysse et al. 2007).

In ALUAM, decisions at the parcel, farm and regional level are optimized in such a way that aggregated land rent is maximized. At the parcel level, decisions concern land-use activities such as crop production, grassland intensity or timber production. Farm-level decisions relate to animal husbandry and flows of fodder and nutrients. Land use links the parcel level with livestock activities through fodder and nutrient balances. At the regional level, resources such as hirable workforce and number of animals available for grazing on summer pastures are only available to a limited extent and are therefore balanced over the whole region. Resource demands of activities at lower levels have to be evaluated to calculate these regional balances. A comparative static approach was used to assess the impact of our four scenarios in 2050 and 2080. We would refer to Briner et al. (2012, 2013a) for details regarding validation and sensitivity analysis of the model simulation.

A crop simulation model was used to integrate information concerning the impact of climate change on agricultural crop yields into ALUAM. This model calculates future yields of relevant crops using data on optimal and absolute crop growing conditions from the Food and Agriculture Organization (FAO) of the United Nations (FAO 2007). We fit a relative crop yield curve for temperature and precipitation values using an incomplete beta distribution. These species-specific crop yield curves serve to calculate the relative yield of crops and grassland for each landscape cell (100 × 100 m) in the case study landscape based on monthly precipitation and temperature values.

Transdisciplinary Process

The modular structure of ALUAM also allowed us to integrate local actors' decision-making into the optimization process using agent-based modeling techniques (Huber et al. 2013a; Hirschi et al. 2013). The use of agents with different preferences and attitudes towards agricultural activities meant we could integrate our findings from the transdisciplinary-based analysis of local decision-making into the model chain. The transdisciplinary process was set up as a cross-sectoral activity to provide guidance for the research process (Seidl et al. 2013). The stakeholder dialogue comprised two strongly inter-linked phases. In the first phase, a steering group representing the core stakeholders was set up in each study region. They consisted of six representatives of typical stakeholder groups who advised the project team from a "study region perspective." In the second phase, regional "transition scenarios" with extended stakeholder groups (10 to 15 representatives) were developed, using a functional-dynamic concept of stakeholder involvement. In four workshops, the research team worked closely with residents and policy-makers in the case study region to develop six local scenarios (Brand et al. 2013). However, in this article we refrain from applying agent-based techniques and use comparative static results. Given the long timeframe considered in our simulations (2080), the use of existing preferences is implausible.

Policy Assessment

Finally, the model chain also allows for the integration of the knowledge on relevant policy issues studied (orange fields in Figure 2). We developed a policy-oriented indicator system for agricultural ecosystem services that can be linked to model-based scenario assessments (Hirschi et al. 2013). Specifically, we used policy network analysis to investigate the structural configuration of the relevant policy subsystems and analyzed the policy preferences of the main political actors in those policy subsystems to assess the feasibility of different policy options in the political process (Ingold et al. 2010; Hirschi 2010). This approach permits us to integrate policy-mediated feedback effects in our model chain. If the modeling outcome deviates greatly from the normative state of the corresponding indicator (i.e., derived from political goals as formulated in official policy documents of the federal government, and not from quantitative valuations), a policy change seems indicated.

Indicators of Ecosystem Services

Food Provision

A wheat-equivalent index was calculated to assess food production. It considered the different values of crops and grassland for human nutrition, recognizing that grassland has a lower human nutritional value because grass must first be converted into milk or meat by animals. Crops that can be digested directly by humans, such as wheat, have a higher nutritional value than grassland. For a detailed description of the index, see Briner et al. (2012). During the optimization process, food provision was calculated as a direct function of the number of animals and the proportion of land in different land-use categories. Food production is impacted by climate change through the spatially explicit yield data provided by the crop model on which the optimization process is based.

Timber Production

Potential timber production within each landscape cell was calculated using the LandClim model (Briner et al. 2012; Temperli et al. 2012). We implemented a forest management regime whereby forest stands are evaluated every 20 years to determine if the stand is ready to be entered and timber removed. If the average height of the dominant trees within a stand (largest 100 trees ha-1) is greater than 15 m, all trees with a diameter at breast height (DBH) greater than 20 cm are harvested. On average, harvested trees had a DBH between 25 and 30 cm. This management routine was used to obtain a potential timber production value for each cell in the landscape. The actual value of timber harvested was calculated in the ALUAM model, whereby production costs and timber prices were also taken into consideration.

Natural Hazard Protection

The ability of forests to provide protection against gravitational hazards, such as rock fall, depends on the location of forest stands as well as on the tree species mix, the structural profile, the rooting stability of live trees and the regeneration potential of the forest. We developed a metric that assesses the protection that forests provide against rock fall (Rockfall Protection Index RFPI; for details see Briner et al. 2013b). Rock fall protection was calculated using simulated forest data from LandClim. We then transferred the spatially explicit protection data to ALUAM, where the optimization process aggregated the potential RFPI of all parcels used as forests or fallow land to calculate the regional protection index.

Carbon Sequestration/Release

In our simulations, net greenhouse gas emissions are influenced by the amount of greenhouse gas (GHG) emitted by agriculture and the carbon sequestered (or emitted) by forests. In general, animal-based food production increases greenhouse gas emissions, and forest growth and timber production enhance sequestration. We accounted for all GHG emissions from agricultural activities, including indirect N2O emissions associated with N losses. The amount of greenhouse gases emitted depends primarily on the number of livestock kept in the region. Land use also has an impact on greenhouse gas emissions, as the number of tractor hours on extensively used grassland is lower than on intensively used grassland, and CO2 emissions are lower as well. In addition, nitrogen throughput is lower on extensively used land, causing lower nitrous oxide emissions. Carbon sequestration by forests was calculated as the amount of carbon immobilized each year in aboveground tree biomass. We assume that forest products lead to a displacement of fossil fuel–intensive products in housing construction and, thus, to a significant reduction in atmospheric carbon if timber is harvested (Perez-Garcia et al. 2005). Only emissions inside the region were considered, and storage as soil organic carbon (e.g., Freibauer et al. 2004; Lal 2005) was not taken into account

Scenarios

In Mountland, the global IPCC SRES scenarios regarding climatic, market and policy developments were regionally downscaled to produce case study–specific storylines that represent local conditions while remaining consistent with the global projections (Walz et al. 2013). This supports the view that global development and local land management are interdependent (Pahl-Wostl 2009). We used these storylines to adapt the economic scenarios provided by Abildtrup et al. (2006) to meet Swiss conditions. This allowed us to anchor the "context scenarios" in the overall climatic and socio-economic developments on a European level since this is most relevant for Swiss agricultural market developments. The scenarios provided in Abildtrup et al. (2006) provide a detailed and comprehensive development of prices and costs for different agricultural activities under the four global change scenarios. Table 2 summarizes the assumptions underlying the regional context scenarios (for details regarding driving forces, e.g., productivity changes and production resources, we refer to Walz et al. 2013).

Table 2: Context scenarios" for the case study region Visp

	Growth and convergence	**Regional centers**	**Green growth**	**Local sustainability**
Climate	A1FI	A2	B1	B2
Consumption patterns	Global production	Regional products	Certified products	Regional products
Agricultural markets	Large decline in prices (open markets)	Stable prices (border protection)	Decline in prices (more open markets)	Stable prices (open markets but increase in EU prices)
Wood prices	Stable prices	Stable prices	Increase in prices	Increase in prices
Technological innovation in agriculture	High innovation rate	Low innovation rate	High innovation rate	Low innovation rate
Nature conservation	Reduction	Reduction	Extension	Extension
Agricultural policy	Liberalization	Reduced domestic support	Greening (high domestic support)	Greening (high domestic support)

Huber et al.

Huber et al. Ecological Processes 2014 3:9, doi: 10.1186/2192-1709-3-9

The regional context scenario Growth and convergence is based on the SRES scenario A1FI, which implies rapid economic growth and an increase in average temperature of 4°C by 2080. Global production processes put pressure on agricultural prices in Switzerland (with a high border protection and high domestic support—e.g., direct payments—for the agricultural sector). Agricultural policies are driven by market liberalization and a decrease in domestic support. To increase productivity, technological innovation in agriculture is high. Wood prices are assumed to remain constant. Global development in the scenario Regional centers is characterized by a growing focus on self-sufficiency and preservation of local identities. The A2 scenario of the IPCC is assumed to be the underlying climate change scenario. The heterogeneous world in this scenario results in a more regional consumption pattern. Innovation rates in agriculture are rather low, and agricultural farm gate prices remain constant. Nature conservation policies decline, and national agricultural policy is dominated by protective measures (i.e., tariffs). The Green growth scenario is characterized by more nature conservation policies and a greening of agricultural policy. Agricultural markets are less protected than today, but consumption patterns focus on certified products representing higher environmental standards. At the same time, innovation rates in agriculture are high. Despite more open market policies, there is only a slight decrease in agricultural farm gate prices, and the price of timber rises. The climate projection is based on the B1 scenario of the IPCC. Local sustainability summarizes a context scenario which combines the B2 scenario with regional consumption patterns. The agricultural sector is characterized by low innovation rates and a decline in farm gate prices. Policies focus on more greening in the agricultural sector and more nature conservation. Timber prices are assumed to rise.A summary of the parameter changes in ALUAM is presented in (Table A1). Lastly, it is noteworthy that an increase in world market prices does not automatically translate into increasing prices on Swiss agricultural markets. Support for agricultural production in Switzerland still rates as one of the highest worldwide, and farm gate prices for milk and meat are over 40% higher compared to the European level (Huber and Lehmann 2010).

RESULTS

Simulation results for relative changes in EGS provision for the years 2050 and 2080 are presented for food and timber provision (Figure

3) and for carbon sequestration and rock fall protection (Figure 4). To reflect the different climatic and ecological conditions in the case study region, we distinguish between low elevations in the main valley around Visp (400–800 m a.s.l.), intermediate elevations (800–1,600 m a.s.l.) and high elevations (above 1,600 m a.s.l.). The scenarios are laid out according to the gradient with respect to the transitions from moderate to severe climate change impacts (B1, B2, A1, A2).

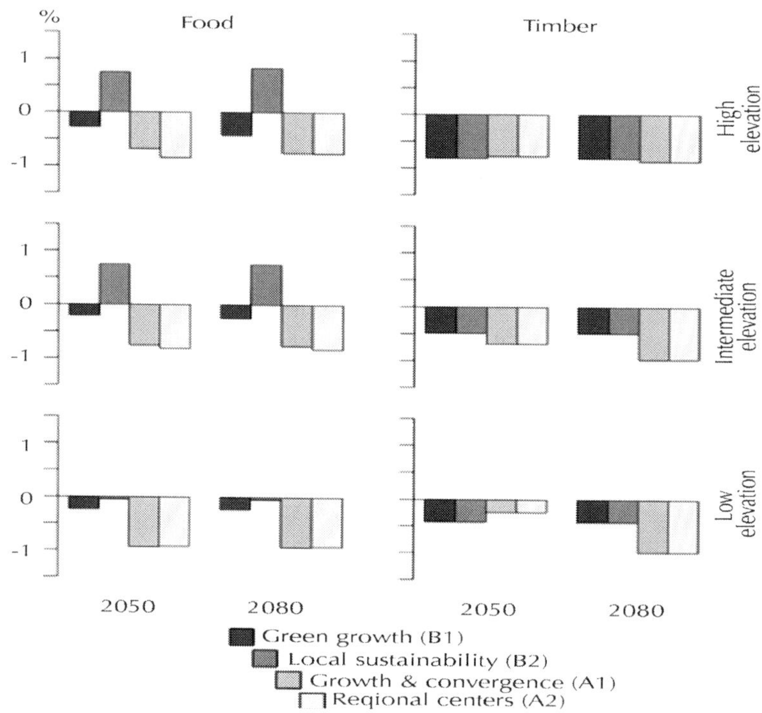

Figure 3: Simulation results for relative changes in food production and timber harvest across three elevation bands (low, intermediate, high) for four global change scenarios in Visp. Changes are always compared to year 2010.

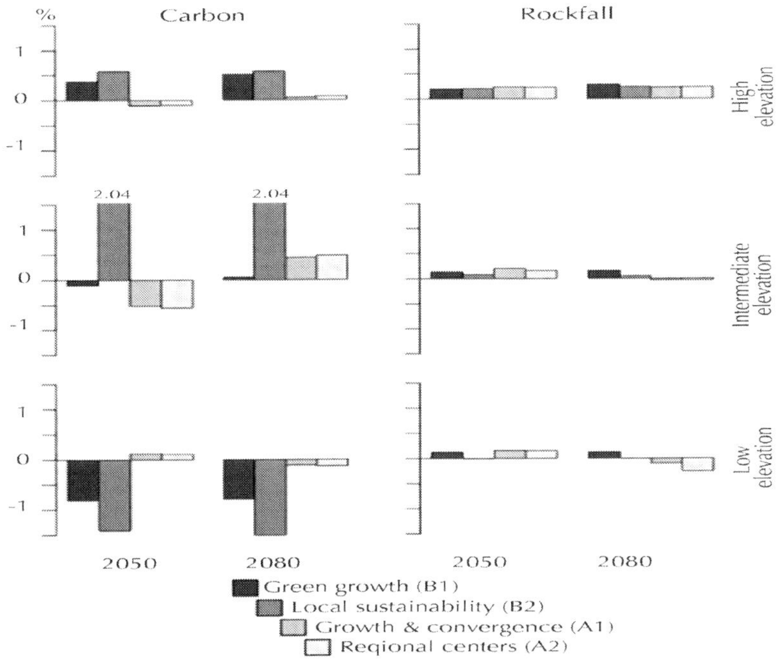

Figure 4: Simulation results for relative changes in net greenhouse gas emissions and rock fall protection across three elevation bands (low, intermediate, high) for four global change scenarios in Visp. Changes are always compared to year 2010.

Food Provision

Food production shows high variability between the families of A and B scenarios as well as between the two B scenarios. There are no appreciable differences between the two time steps 2050 and 2080 for any of the elevations. Specifically, food production declines strongly in the A scenarios. In the scenario Growth and convergence (A1), low prices for food and a reduction in agricultural support are projected to result in low intensity farming based mainly on low input meat and milk production. Two developments drive this drop in food provision. At low elevations, a shift from crop to milk production leads to a significant decline in food provision due to the lower human nutritional value of grassland compared to arable land. At higher elevations, simulation results show a decrease in the number of animals and land abandonment at marginal agronomic sites (for details of this scenario, see Briner et al. 2012). The results of the scenario Regional centers (A2) are based on the same two mechanisms. In contrast to the Growth and convergence (A1) scenario, however, these effects are not due to more open markets, but to a combination of low domestic support (i.e., direct payments) and an increase in costs which are associated with a more heterogeneous world in the A2 scenario. Food production is also lower in the Green growth (B1) scenario. However, some of the losses in economic returns are compensated by an increase in ecologically oriented direct payments. In the last scenario, Local sustainability (B2), these additional payments are combined with constant prices based on an increased level of European production prices. In this case, simulation results imply the reverse of the effects seen in the other scenarios. First, food production does not decline at lower elevations since arable land is used for crop production. Second, there is an increase in the number of animals kept for meat production (suckler cows and sheep) at intermediate and high elevations.

Timber Production

Climate change is projected to have a differential impact on potential timber production along the elevational gradient and through time (Figure A1 in At lower sites, drought leads to a decline in potential timber production and a shift towards more drought-tolerant species. The reduction is accentuated for more severe climate scenarios (e.g., Growth and convergence [A1] and Regional centers [A2]) after 2050 when drought years become more frequent. In contrast, at intermediate and high elevations climate change is projected to increase potential timber production. However, the ALUAM simulation of the actual harvest, which also takes prices and costs for different types of wood into account, indicates that timber production declines in all scenarios and at all elevations (Figure 3). There are two reasons for this development. First, an increase in energy prices (i.e., fuel) makes the production of timber unprofitable on steep terrain even given a slight rise in prices under the scenarios Green growth (B1) and Local sustainability (B2). Second, drought-induced tree shifts lead to a scarcity of more profitable tree species, mainly at low and intermediate elevations. At low elevations, these effects are combined with a simulated reduction in potential timber production which is particularly accentuated in the scenarios Growth and convergence (A1) and Regional centers (A2) in 2080. The increase in potential timber production at high elevations is dominated by Picea abies, which is found to be less profitable in our simulations compared to the current market situation. In addition to the direct impacts of changes in climate and market forces, forest EGS provision at high elevations is affected by land-use change, i.e., when agricultural land is abandoned. However, the direct impact of climate and wood market changes are projected to have a greater effect on the state of the forest, and thus on timber production, than land-use changes (Briner et al. 2013a).

Natural Hazard Protection

In our case study region, climate change influences not only timber production but also the protection that forests provide against natural hazards such as rock fall. Thus, changes in natural hazard protection reflect general forest development, and this is dominated by a drought-induced decline in biomass and a shift towards more drought-tolerant tree species at low elevations and an increase in biomass production at higher elevations. Figure 5 illustrates the spatial heterogeneity in changes of the projected rock fall protection index by 2080 under the four scenarios. Rock fall protection decreases at lower elevations under the more severe climate scenarios and remains above the 2010 level only in the Green growth (B1) scenario. In contrast, protection from rock fall increases at high elevations in all scenarios. At intermediate elevations, rock fall protection declines slightly by 2080 in the scenarios based on more severe climate change, such as Growth and convergence (A1) and Regional centers (A2) which is illustrated by the aggregated indicator in Figure 4.

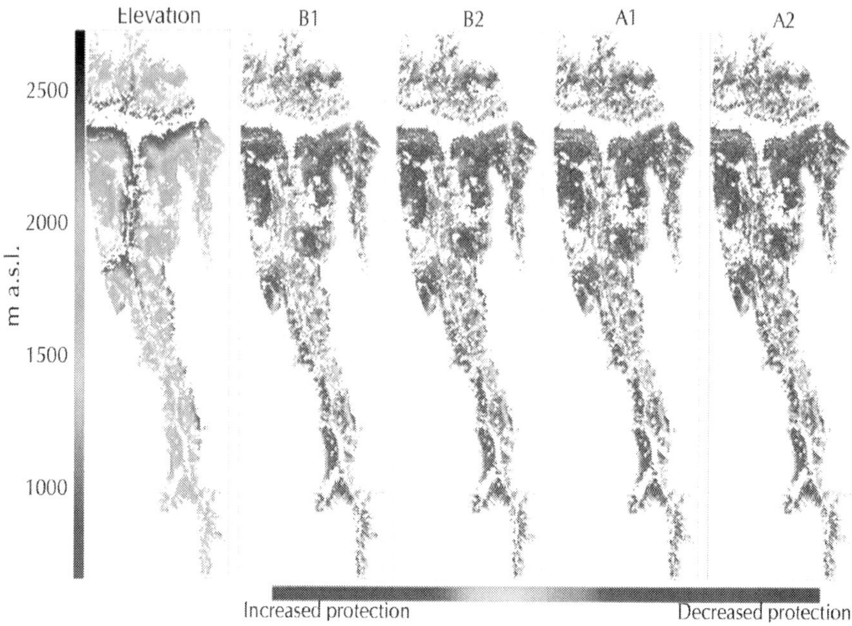

Figure 5: Spatially explicit provision of rock fall protection. The four context scenarios Green growth (B1), Local sustainability (B2), Growth and convergence (A1), and Regional centers (A2) are presented for the case study region Visp.

Carbon Sequestration/Release

Results in Figure 4 imply that greenhouse gas sequestration and release will differ significantly across scenarios and elevational gradients and over time. Projected increases in crop production and reductions in livestock numbers at lower elevations result in lower greenhouse gas emissions by agriculture in the scenarios Green growth (B1) and Local sustainability (B2) by 2050. However, simulation results for the Growth and convergence (A1) and Regional centers (A2) scenarios indicate a shift from crop to milk production and, thus, a slight increase in greenhouse gas emissions. The extensification of animal production is projected to reduce greenhouse gas emissions by 2080 in these scenarios. These effects over-compensate the lower carbon sequestration by forests arising from the decline in timber production and the drought-induced reduction in forest biomass. The increase in the number of animals at intermediate elevations in the Local sustainability (B2) scenario results in a large rise in greenhouse gas emissions compared to the 2010 level. In the other scenarios, however, the decline of timber harvesting and the corresponding loss of carbon sequestration outweigh the reduced greenhouse gas emission induced by a lower number of animals for the year 2080. This effect also leads to changes in net carbon sequestration at high elevations. The substantial reduction in timber harvest decreases net greenhouse gas sequestration for all scenarios by 2080. Thus, the effect of reduced greenhouse gas emissions induced by a lower level of food provision in the Growth and convergence (A1), Regional centers (A2) and Green growth (B1) scenarios is not evident in the net balance of greenhouse gas emissions and carbon sequestration.

DISCUSSION

Discussion of Scenario Analysis

Simulation results illustrate the effect of global change impacts on EGS in a mountain case study region based on regional socio-economic and climate scenarios that are consistent with global IPCC SRES scenarios. Overall, given the different scenarios, the variance between the expected effects is high, even within IPCC SRES scenarios from the same family, such as Green growth (B1) and Local sustainability (B2).The key findings can be summarized in four points. First, the scenario analysis revealed the spatial and temporal heterogeneity of EGS provision, even at a rather small (350 km2) regional scale. In this inner-alpine case study region, we found that EGS are highly vulnerable at all elevations, depending on the rate of climate change and the magnitude of socio-economic changes. Second, simulation results indicate that the impact of climatic change is much more pronounced for forest EGS compared to agricultural EGS. This reflects our findings that at low elevations, increasing drought is projected to reduce both forest biomass and provoke a switch towards more drought-tolerant tree species (Elkin et al. 2013). These results correspond with our experimental research in the same area, which provided important data for modeling drought and management effects for different tree species (Eilmann and Rigling 2012; Eilmann et al. 2009; Rigling et al. 2013). In contrast, changes to agricultural EGS were found to be due primarily to shifts in economic conditions that alter land use and land management (Briner et al. 2012, 2013a). Third, our simulation results suggest that trade-offs between different EGS will differ under the different scenarios, such as between food provision at higher elevations and increased agricultural greenhouse gas emissions. These trade-offs depend on the complex interactions of underlying structural and environmental conditions that are driven by socio-economic and climatic factors (Briner et al. 2013b). The emergence of trade-offs in an integrated modeling framework allowed us to address the economic aspects of marginality and substitution effects which are important in the valuation of EGS (e.g., Fisher et al. 2008; Bateman et al. 2011; Farley 2012). Fourth, our results demonstrate that land use and the associated provision of EGS are not only influenced directly by environmental shifts and economic decisions, but also depend strongly on the interactions between these two components (Briner et al. 2012). As has been emphasized by many other authors (e.g., de Groot et al. 2002, 2010; Haines- Young and Potschin 2009; Carpenter et al. 2009), this latter aspect demonstrates the importance of including ecological processes in socio-economic modeling frameworks for the assessment of EGS. Ecological processes, such as the development of tree species under drought conditions or the variance in agricultural crop and grassland yields at higher temperatures and different elevational gradients, form a fundamental basis for the quantitative assessment of EGS changes. This would not have been possible in individual studies addressing forestry and agricultural activities in isolated socio-economic or ecological frameworks without feedbacks. This has also been demonstrated in other modeling studies addressing erosion (Claessens et al. 2009), hydrological cycles (Bithell and Brasington 2009), natural reforestation (Gibon et al. 2010) and habitat conservation (Huber et al. 2013a). Spatially explicit modeling across scales and disciplines requires compromises, and there are several methodological aspects that should be considered within this context. The indicators that were used in our approach to link agricultural and forest activities to EGS provision, for instance, are more general than we would have liked, and indicators that allow for a more comprehensive illustration of the feedbacks between EGS would certainly improve our modeling results (e.g., van Oudenhoven et al. 2012; Müller and Burkhard 2012). In addition, the alignment of the time scales between the forest simulation model, which focuses on long-term developments, and the socio-economic land-use model, which has limited power in long-term projections (i.e., comparative static modeling approaches) remains a challenge (see Rounsevell and Arneth 2011). Furthermore, our spatially explicit approach allowed for landscape-level simulations that include relevant spatial dependencies on different scales (from plot-based observational and experimental findings to the region). However, we did not consider off-site effects (i.e., effects outside of the region being investigated), albeit they are an important facet in EGS analysis especially with respect to carbon sequestration (see Seppelt et al. 2011).

Integrated Modeling Results in the Context of Policy Processes and Stakeholder Interactions

Our modeling results emphasize the importance of heterogeneity, trade-offs and feedbacks in the assessment of future EGS provision in mountain regions (see also Huber et al. 2013c). However, quantitative scenario analysis will not suffice to bring EGS into policy and real decision-making processes, since societal values and political trade-offs must also be taken into account (Daily et al. 2009). In this context, taking into account policy processes and the consideration of stakeholders who actually manage resources in the respective area is a crucial aspect for EGS studies (e.g., Cowling et al. 2008; Daily and Matson 2008; Reed et al. 2009). Our research approach considered these factors by analyzing the policy network of central land use policies as well as by developing regional scenarios for the Visp area in collaboration with key stakeholders who are closely linked to the context scenarios presented here. We defined six consistent, multi-scale scenarios covering the global, the Swiss and the regional levels (i.e., the combination of the "context scenarios" presented here with local scenarios developed by the stakeholders). The scenarios represent possible futures and indicate strategies to cope with local and global challenges (Brand et al. 2013).

The results from the scenario development have two implications. First, when confronted with the consequences of global change impacts, local stakeholders became aware of a wider systematic and realistic picture. In our local scenario workshops, for example, stakeholders did not consider two out of the six multi-scale scenarios at the outset of the transdisciplinary process since the critical combination of climatic and economic risks in the region were implicitly avoided in the stakeholder discussions (Brand et al. 2013). Thus, the consideration of ecological processes and the use of mechanistic ecological models to produce quantitative projections of future states in transdisciplinary approaches turned out to be a useful means of confronting stakeholders with unwelcome, but nevertheless possible, prospects for the future rather than defining scenarios based on wishful thinking. Second, the results from the interdisciplinary process reveal that two impact factors (i.e., spatial planning and budget) represent the most important control factors for strategies addressing the challenges generated by global change in mountain regions. More importantly, local identity and the affinity local residents have to the region, its products and traditions, emerged as the most important, but also most ambivalent, elements of the social-ecological system analyzed. Thus, traditional land-use activities will play a crucial and yet unpredictable role in the future development of the Visp region.

In addition to stakeholder integration, the incorporation of more specific knowledge regarding the existing political structures and policy variables helped to make model-based ex-ante assessments of the future provision of EGS politically more realistic and, thus, more useful and relevant for actual policy-making processes (Hirschi et al. 2013). Within the Mountland project, we analyzed the political network structure of local infrastructure projects (Ingold et al. 2010) and the governance modes of regional adaptation measures, as well as the political dynamics and policy preferences in national agricultural and forest policy networks (Hirschi et al. 2013). The analyses revealed two central elements of relevant policy processes to effectively address the vulnerability of EGS in mountain regions. First, the promotion of integrative approaches, i.e., increased coordination of sectoral policies, and second, a more network-oriented management and steering of political processes with better integration of local stakeholders (Huber et al. 2013c).

However, our findings also reveal two important limitations that hinder the integration of EGS assessments, and thus a better consideration of ecological processes in policy-making. For instance, Ingold et al. (2010) showed that existing local policy projects give high consideration to regional planning as well as communal and economic interests, while most of the actors defending ecological concerns are

not well embedded in the relevant information-exchange networks. Since there is a high degree of uncertainty, informal processes are most likely to gain increasing importance in the context of adaptation to climate change. In addition, our analysis of agricultural and forest policy networks showed a strong path-dependence and sectoral orientation of these two important land-use policies. This, combined with the dominance of economic interests, particularly in the agricultural sector, make major policy reforms in both sectors quite unlikely at present (Hirschi et al. 2013).

CONCLUSIONS

We have described a framework that effectively aids the integration of ecological processes based on experiments and observations into an integrative modeling chain of EGS provision in mountain regions, the political decision-making process and also transdisciplinary stakeholder interactions. Simulation results illustrated (i) the variance between regionally downscaled SRES scenarios, (ii) the different drivers for the future provision of EGS from forest and agriculture in the different scenarios, (iii) the different trade-offs in the four simulated scenarios and (iv) the importance of the consideration of social-ecological systems interactions. The discussion of these results in the context of policy processes and stakeholder interactions shows the usefulness of quantitative projections of future EGS anchored in the analysis of ecological processes in transdisciplinary approaches. It also reveals existing enabling factors and limitations that should be considered for a better integration of ecological processes in actual policy-making that wishes to address EGS provision.

AUTHORS' CONTRIBUTIONS

HB and AR designed the research; SB, CE and RSe performed simulation analyses, CH performed policy analysis, RS performed transdisciplinary research; RH wrote the manuscript; SB, CE, RSe, CH, RSn, HB and AR reviewed and commented on various versions of the manuscript. All authors read and approved the final manuscript.

ACKNOWLEDGEMENTS

This work was supported by the CCES (Competence Centre Environment and Sustainability of the ETH Domain, Switzerland) as part of the Mountland project. We thank Jenny Bays for language corrections and the two anonymous reviewers, as well as the editors, for their constructive and helpful comments on earlier versions of the manuscript.

REFERENCES

1. Abildtrup J, Audsley E, Fekete-Farkas M, Giupponi C, Gylling M, Rosato P, Rounsevell M (2006) Socio-economic scenario development for the assessment of climate change impacts on agricultural land use: a pairwise comparison approach. Environ Sci Pol 9(2):101.
2. Bateman I, Mace G, Fezzi C, Atkinson G, Turner K (2011) Economic analysis for ecosystem service assessments. Environ Resour Econ 48(2):177-218.
3. Bithell M, Brasington J (2009) Coupling agent-based models of subsistence farming with individual-based forest models and dynamic models of water distribution. Environ Model Softw 24(2):173-190
4. Braat LC, de Groot R (2012) The ecosystem services agenda: bridging the worlds of natural science and economics, conservation and development, and public and private policy. Ecosyst Serv 1(1):4-15.
5. Brand F, Seidl R, Le QB, Brändle J, Scholz RW (2013) Constructing consistent multiscale scenarios by transdisciplinary processes: the case of mountain regions facing global change. Ecol Soc 18(2):43
6. Briner S, Huber R, Elkin C, Grêt-Regamey A (2012) Assessing the impacts of economic and climate changes on land-use in mountain regions: a spatial dynamic modeling approach. Agr Ecosyst Environ 149(2012):50-63
7. Briner S, Elkin C, Huber R (2013a) Evaluating the relative impact of climate and economic changes on forest and

agricultural ecosystem services in mountain regions. J Environ Manage 129(0):414-422 doi:http://dx.doi.org/10.1016/j.jenvman.2013.07.018

8. Briner S, Huber R, Bebi P, Elkin C, Schmatz DR, Grêt-Regamey A (2013b) Trade-offs between ecosystem services in a mountain region. Ecol Soc 18(3):
9. Brunner I, Pannatier EG, Frey B, Rigling A, Landolt W, Zimmermann S, Dobbertin M (2009) Morphological and physiological responses of Scots pine fine roots to water supply in a dry climatic region in Switzerland. Tree Physiol 29(4):541-550 doi:10.1093/treephys/tpn046
10. Buysse J, van Huylenbroeck G, Lauwers L (2007) Normative, positive and econometric mathematical programming as tools for incorporation of multifunctionality in agricultural policy modelling. Agr Ecosyst Environ 120:70-81
11. Carpenter SR, Mooney HA, Agard J, Capistrano D, DeFries RS, Diaz S, Dietz T, Duraiappah AK, Oteng-Yeboah A, Pereira HM, Perrings C, Reid WV, Sarukhan J, Scholes RJ, Whyte A (2009) Science for managing ecosystem services: beyond the Millennium Ecosystem Assessment. Proc Natl Acad Sci USA 106(5):1305-1312
12. Cash DW, Adger WN, Berkes F, Garden P, Lebel L, Olsson P, Pritchard L, Young O (2006) Scale and cross-scale dynamics: governance and information in a multilevel world. Ecol Soc 11(2):8
13. Chapin FS, Lovecraft AL, Zavaleta ES, Nelson J, Robards MD, Kofinas GP, Trainor SF, Peterson GD, Huntington HP, Naylor RL (2006) Policy strategies to address sustainability of Alaskan boreal forests in response to a directionally changing climate. Proc Natl Acad Sci USA 103(45):16637-16643
14. Claessens L, Schoorl JM, Verburg PH, Geraedts L, Veldkamp A (2009) Modelling interactions and feedback mechanisms between land use change and landscape processes. Agric Ecosyst Environ 129(1–3):157-170
15. Collins SL, Carpenter SR, Swinton SM, Orenstein DE, Childers DL, Gragson TL, Grimm NB, Grove JM, Harlan SL, Kaye JP, Knapp AK, Kofinas GP, Magnuson JJ, McDowell WH, Melack JM, Ogden LA, Robertson GP, Smith MD, Whitmer AC (2011) An integrated

conceptual framework for long-term social-ecological research. Front Ecol Environ 9(6):351-357

16. Cowling RM, Egoh B, Knight AT, O'Farrell PJ, Reyers B, Rouget M, Roux DJ, Welz A, Wilhelm-Rechman A (2008) An operational model for mainstreaming ecosystem services for implementation. Proc Natl Acad Sci USA 105(28):9483-9488
17. Cumming G, Alcamo J, Sala O, Swart R, Bennett E, Zurek M (2005) Are existing global scenarios consistent with ecological feedbacks? Ecosystems 8(2):143-152
18. Daily GC, Matson PA (2008) Ecosystem services: from theory to implementation. Proc Natl Acad Sci USA 105(28):9455-9456
19. Daily GC, Polasky S, Goldstein J, Kareiva PM, Mooney HA, Pejchar L, Ricketts TH, Salzman J, Shallenberger R (2009) Ecosystem services in decision making: time to deliver. Front Ecol Environ 7(1):21-28
20. de Groot RS, Wilson MA, Boumans RMJ (2002) A typology for the classification, description and valuation of ecosystem functions, goods and services. Ecol Econ 41(3):393-408
21. de Groot RS, Alkemade R, Braat L, Hein L, Willemen L (2010) Challenges in integrating the concept of ecosystem services and values in landscape planning, management and decision making. Ecol Complex 7(3):260-272
22. de Groot R, Brander L, van der Ploeg S, Costanza R, Bernard F, Braat L, Christie M, Crossman N, Ghermandi A, Hein L, Hussain S, Kumar P, McVittie A, Portela R, Rodriguez LC, ten Brink P, van Beukering P (2012) Global estimates of the value of ecosystems and their services in monetary units. Ecosyst Serv 1(1):50-61
23. Diaz S, Quétier F, Caceres DM, Trainor SF, Pérez-Harguindeguy N, Bret-Harte MS, Finegan B, Pena-Claros M, Poorter L (2011) Linking functional diversity and social actor strategies in a framework for interdisciplinary analysis of nature's benefits to society. Proc Natl Acad Sci USA 108(3):895-902
24. Dobbertin M, Eilmann B, Bleuler P, Giuggiola A, Graf Pannatier E, Landolt W, Schleppi P, Rigling A (2010) Effect of irrigation on needle morphology, shoot and stem growth in a drought-exposed Pinus sylvestris forest. Tree Physiol 30(3):346-360
25. FAO (Food and Agriculture Organization of the UN) (2007) Ecocrop. http://ecocrop.fao.org/ecocrop/srv/en/home

26. Eilmann B, Rigling A (2012) Tree-growth analyses to estimate tree species' drought tolerance. Tree Physiol 32(2):178-187
27. Eilmann B, Zweifel R, Buchmann N, Fonti P, Rigling A (2009) Drought-induced adaptation of the xylem in Scots pine and pubescent oak. Tree Physiol 29(8):1011-1020
28. Eilmann B, Buchmann N, Siegwolf R, Saurer M, Cherubini P, Rigling A (2010) Fast response of Scots pine to improved water availability reflected in tree-ring width and δ13C. Plant Cell Environ 33(8):1351-1360
29. Eilmann B, Zweifel R, Buchmann N, Graf Pannatier E, Rigling A (2011) Drought alters timing, quantity, and quality of wood formation in Scots pine. J Exp Bot 62(8):2763-2771
30. Eilmann B, Dobbertin M, Rigling A (2013) Growth response of Scots pine with different crown transparency status to drought release. Ann For Sci 70(7):685-693
31. Elkin C, Bugmann H (2010) Impact of climate change on ecosystem services in the Valais, Switzerland. In: Price M (ed) Europe's ecological backbone: recognising the true value of our mountains, European Environment Agency, Copenhagen. pp 67-70
32. Elkin C, Reineking B, Bigler C, Bugmann H (2012) Do small-grain processes matter for landscape scale questions? Sensitivity of a forest landscape model to the formulation of tree growth rate. Landsc Ecol 27(5):697-711
33. Elkin C, Gutiérrez AG, Leuzinger S, Manusch C, Temperli C, Rasche L, Bugmann H (2013) A 2°C warmer world is not safe for ecosystem services in the European Alps. Glob Chang Biol 19(6):1827-1840
34. Euliss NH, Smith LM, Liu S, Duffy WG, Faulkner SP, Gleason RA, Eckles SD (2011) Integrating estimates of ecosystem services from conservation programs and practices into models for decision makers. Ecol Appl 21(sp1):S128-S134
35. Farley J (2012) Ecosystem services: the economics debate. Ecosyst Serv 1(1):40-49
36. Fisher B, Turner K, Zylstra M, Brouwer R, Groot R, Farber S, Ferraro P, Green R, Hadley D, Harlow J, Jefferiss P, Kirkby C, Morling P, Mowatt S, Naidoo R, Paavola J, Strassburg B, Yu D, Balmford A

(2008) Ecosystem services and economic theory: integration for policy-relevant research. Ecol Appl 18(8):2050-2067

37. Freibauer A, Rounsevell MDA, Smith P, Verhagen J (2004) Carbon sequestration in the agricultural soils of Europe. Geoderma 122(1):1-23
38. Gibon A, Sheeren D, Monteil C, Ladet S, Balent G (2010) Modelling and simulating change in reforesting mountain landscapes using a social-ecological framework. Landsc Ecol 25(2):267-285
39. Gómez-Baggethun E, de Groot R, Lomas PL, Montes C (2010) The history of ecosystem services in economic theory and practice: from early notions to markets and payment schemes. Ecol Econ 69(6):1209-1218
40. Haines-Young RH, Potschin MB (2009) Methodologies for defining and assessing ecosystem services. Final report, JNCC:69, project code C08-0170-0062. Centre for Environmental Management, Nottingham.
41. Hein L, van Koppen K, de Groot RS, van Ierland EC (2006) Spatial scales, stakeholders and the valuation of ecosystem services. Ecol Econ 57(2):209-228
42. Heiniger U, Theile F, Rigling A, Rigling D (2011) Blue-stain infections in roots, stems and branches of declining Pinus sylvestris trees in a dry inner alpine valley in Switzerland. For Pathol 41(6):501-509
43. Hirschi C (2010) Strengthening regional cohesion: collaborative networks and sustainable development in Swiss rural areas. Ecol Soc 115(4):16
44. Hirschi C, Widmer A, Briner S, Huber R (2013) Combining policy network and model-based scenario analyses: an assessment of future ecosystem goods and services in Swiss mountain regions. Ecol Soc 18(2):
45. Huber R, Lehmann B (2010) WTO agreement on agriculture: potential consequences for agricultural production and land-use patterns in the Swiss lowlands. Dan J Geogr 109(2):131-145
46. Huber R, Briner S, Peringer A, Lauber S, Seidl R, Widmer A, Gillet F, Buttler A, Le QB, Hirschi C (2013a) Modeling social-ecological feedback effects in the implementation of payments for environmental services in pasture-woodlands. Ecol Soc 18(2):

47. Huber R, Bugmann H, Buttler A, Rigling A (2013b) Sustainable land-use practices in European mountain regions under global change: an integrated research approach. Ecol Soc 18(3):

48. Huber R, Rigling A, Bebi P, Brand FS, Briner S, Buttler A, Elkin C, Gillet F, Grêt-Regamey A, Hirschi C, Lischke H, Scholz RW, Seidl R, Spiegelberger T, Walz A, Zimmermann W, Bugmann H (2013c) Sustainable land use in mountain regions under global change: synthesis across scales and disciplines. Ecol Soc 18(3)

49. Ingold K, Balsiger J, Hirschi C (2010) Climate change in mountain regions: how local communities adapt to extreme events. Local Environ 15(7):651-661

50. doi:10.1080/13549839.2010.498811

51. IPCC (2000) Emissions scenarios—summary for policymakers. A special report of IPCC Working Group III, Intergovernmental Panel on Climate Change. IPCC, Geneva.

52. Kremen C (2005) Managing ecosystem services: what do we need to know about their ecology? Ecol Lett 8(5):468-479

53. Kremen C, Williams NM, Aizen MA, Gemmill-Herren B, LeBuhn G, Minckley R, Packer L, Potts SG, Ta R, Steffan-Dewenter I, Vázquez DP, Winfree R, Adams L, Crone EE, Greenleaf SS, Keitt TH, Klein A-M, Regetz J, Ricketts TH (2007) Pollination and other ecosystem services produced by mobile organisms: a conceptual framework for the effects of land-use change. Ecol Lett 10(4):299-314

54. Lal R (2005) Forest soils and carbon sequestration. For Ecol Manage 220(1–3):242-258

55. doi:http://dx.doi.org/10.1016/j.foreco.2005.08.015

56. MEA (2005) Millennium Ecosystem Assessment, ecosystems and human well-being: synthesis. Island Press, Washington DC.

57. Müller F, Burkhard B (2012) The indicator side of ecosystem services. Ecosyst Serv 1(1):26-30

58. Nelson E, Mendoza G, Regetz J, Polasky S, Tallis H, Cameron D, Chan KMA, Daily GC, Goldstein J, Kareiva PM, Lonsdorf E, Naidoo R, Ricketts TH, Shaw M (2009) Modeling multiple ecosystem services, biodiversity conservation, commodity production, and tradeoffs at landscape scales. Front Ecol Environ 7(1):4-11

59. Nicholson E, Mace GM, Armsworth PR, Atkinson G, Buckle S, Clements T, Ewers RM, Fa JE, Gardner TA, Gibbons J, Grenyer R, Metcalfe R, Mourato S, Muûls M, Osborn D, Reuman DC, Watson C, Milner-Gulland EJ (2009) Priority research areas for ecosystem services in a changing world. J Appl Ecol 46(6):1139-1144
60. Oteros-Rozas E, Gonzalez J, Martin-Lopez B, Lopez C, Zorrilla-Miras P, Montes C (2012) Evaluating ecosystem services in transhumance cultural landscapesan interdisciplinary and participatory framework. GAIA Ecol Perspect Sci Soc 21(3):185-193
61. Pahl-Wostl C (2009) A conceptual framework for analysing adaptive capacity and multi-level learning processes in resource governance regimes. Glob Environ Chang 19(3):354-365 doi:http://dx.doi.org/10.1016/j.gloenvcha.2009.06.001
62. Perez-Garcia J, Lippke B, Comnick J, Manriquez C (2005) An assessment of carbon pools, storage, and wood products market substitution using life-cycle analysis results. Wood Fiber Sci 37:140-148
63. Polasky S, Segerson K (2009) Integrating ecology and economics in the study of ecosystem services: some lessons learned. Annu Rev Resour Econ 1(1):409-434
64. doi:10.1146/annurev.resource.050708.144110
65. Reed MS, Bonn A, Slee W, Beharry-Borg N, Birch J, Brown I, Burt TP, Chapman D, Chapman PJ, Clay GD, Cornell SJ, Fraser EDG, Glass JH, Holden J, Hodgson JA, Hubacek K, Irvine B, Jin N, Kirkby MJ, Kunin WE, Moore O, Moseley D, Prell C, Price MF, Quinn CH, Redpath S, Reid C, Stagl S, Stringer LC, Termansen M (2009) The future of the uplands. Land Use Pol 26(Supplement 1):S204-S216
66. Reyers B, Roux DJ, O'Farrell PJ (2010) Can ecosystem services lead ecology on a transdisciplinary pathway? Environ Conserv 37(04):501-511
67. doi:10.1017/S0376892910000846
68. Rigling A, Eilmann B, Koechli R, Dobbertin M (2010) Mistletoe-induced crown degradation in Scots pine in a xeric environment. Tree Physiol 30(7):845-852

69. Rigling A, Bigler C, Eilmann B, Feldmeyer-Christe E, Gimmi U, Ginzler C, Graf U, Mayer P, Vacchiano G, Weber P, Wohlgemuth T, Zweifel R, Dobbertin M (2013) Driving factors of a vegetation shift from Scots pine to pubescent oak in dry Alpine forests. Glob Chang Biol 19(1):229-240
70. Rounsevell MDA, Arneth A (2011) Representing human behaviour and decisional processes in land system models as an integral component of the earth system. Glob Environ Chang 21(3):840-843
71. Rounsevell M, Dawson T, Harrison P (2010) A conceptual framework to assess the effects of environmental change on ecosystem services. Biodivers Conserv 19(10):2823-2842 doi:10.1007/s10531-010-9838-5
72. Schaeffer PV (2008) Thoughts concerning the economic valuation of landscapes. J Environ Manage 89(3):146-154
73. Schreinemachers P, Berger T (2011) An agent-based simulation model of human-environment interactions in agricultural systems. Environ Model Softw 26(7):845-859
74. Schumacher S, Bugmann H (2006) The relative importance of climatic effects, wildfires and management for future forest landscape dynamics in the Swiss Alps. Glob Chang Biol 12(8):1435-1450
75. Schumacher S, Bugmann H, Mladenoff DJ (2004) Improving the formulation of tree growth and succession in a spatially explicit landscape model. Ecol Model 180(1):175-194
76. Seidl R, Brand F, Stauffacher M, Krütli P, Le Q, Spörri A, Meylan G, Moser C, González M, Scholz R (2013) Science with society in the anthropocene. AMBIO 42(1):5-12
77. Seppelt R, Dormann CF, Eppink FV, Lautenbach S, Schmidt S (2011) A quantitative review of ecosystem service studies: approaches, shortcomings and the road ahead. J Appl Ecol 48(3):630-636
78. Temperli C, Bugmann HKM, Elkin C (2012) Adaptive management for competing forest goods and services under climate change. Ecol Appl 22:2065-2077 doi:10.1890/12-0210.1
79. van Oudenhoven APE, Petz K, Alkemade R, Hein L, de Groot RS (2012) Framework for systematic indicator selection to assess

effects of land management on ecosystem services. Ecol Indic 21:110-122

80. Wainger L, Mazzotta M (2011) Realizing the potential of ecosystem services: a framework for relating ecological changes to economic benefits. Environ Manage 48(4):710-733
81. Walz A, Braendle JM, Lang DJ, Brand F, Briner S, Elkin C, Hirschi C, Huber R, Lischke H, Schmatz DR (2013) Experience from downscaling IPCC-SRES scenarios to specific national-level focus scenarios for ecosystem service management. Technol Forecast Soc Change.
82. Weber P, Bugmann H, Fonti P, Rigling A (2008) Using a retrospective dynamic competition index to reconstruct forest succession. For Ecol Manage 254(1):96-106
83. Wermelinger B, Rigling A, Schneider Mathis D, Dobbertin M (2008) Assessing the role of bark- and wood-boring insects in the decline of Scots pine (Pinus sylvestris) in the Swiss Rhone valley. Ecol Entomol 33(2):239-249
84. Young O (2002) The institutional dimensions of environmental change: fit, interplay, and scale. MIT Press, Cambridge.
85. Zweifel R, Rigling A, Dobbertin M (2009) Species-specific stomatal response of trees to drought—a link to vegetation dynamics? J Veg Sci 20(3):442-454

Citations

CHAPTER 1

R. Saha, R. S. Chaudhary, and J. Somasundaram, "Soil Health Management under Hill Agroecosystem of North East India," Applied and Environmental Soil Science, vol. 2012, Article ID 696174, 9 pages, 2012. doi:10.1155/2012/696174.

CHAPTER 2

S. Adjei-Nsiah, "Role of Pigeonpea Cultivation on Soil Fertility and Farming System Sustainability in Ghana,"International Journal of Agronomy, vol. 2012, Article ID 702506, 8 pages, 2012. doi:10.1155/2012/702506.

CHAPTER 3

Audun Korsaeth, "N, P, and K Budgets and Changes in Selected Topsoil Nutrients over 10 Years in a Long-Term Experiment with Conventional and Organic Crop Rotations," Applied and Environmental Soil Science, vol. 2012, Article ID 539582, 17 pages, 2012. doi:10.1155/2012/539582.

CHAPTER 4

Subramaniam Gopalakrishna, Vadlamudi Srinivas, Gottumukkala Alekhya,Bandikinda Prakash, Himabindu Kudapa, Abhishek Rathore, and Rajeev Kumar Varshney, The Extent of Grain Yield and Plant Growth Enhancement by Plant Growth-Promoting Broad-Spectrum-*Streptomyces* Sp. in Chickpea, doi:10.1186/s40064-015-0811-3

CHAPTER 5

P. Walter Stinner, the Use of Legumes as a Biogas Substrate - Potentials for Saving Energy and Reducing Greenhouse Gas Emissions through Symbiotic Nitrogen Fixation, doi:10.1186/s13705-015-0034.

CHAPTER 6

Vanesa Rodríguez Osuna, Jan Börner, Udo Nehren, Rachel Bardy Prado, Hartmut Gaese, and Jürgen Heinrich, Priority areas for watershed service conservation in the Guapi-Macacu region of Rio de Janeiro, Atlantic Forest, Brazil, doi: 10.1186/s13717-014-0016-7.

CHAPTER 7

An MO Notenbaert, Jonathan Davies, Jan De Leeuw, Mohammed Said, Mario Herrero, Pablo Manzano, Michael Waithaka, Abdilahi Aboud, and Shadrack Omondi, Policies in Support of Pastoralism and Biodiversity in the Heterogeneous Drylands of East Africa, doi: 10.1186/2041-7136-2-14.

CHAPTER 8

Robert Huber, Simon Briner, Harald Bugmann, Ché Elkin, Christian Hirschi, Roman Seidl, Rebecca Snell, and Andreas Rigling, Inter- and Trans Disciplinary Perspective on the Integration of Ecological Processes into Ecosystem Services Analysis in A Mountain Region, doi:10.1186/2192-1709-3-9.

Index

A

B

C

D

N

O

P

R

S

T

V

W